T0281444

Forum for Interdisciplinary Mathematics

The *Forum for Interdisciplinary Mathematics* series publishes high-quality monographs and lecture notes in mathematics and interdisciplinary areas where mathematics has a fundamental role, such as statistics, operations research, computer science, financial mathematics, industrial mathematics, and bio-mathematics. It reflects the increasing demand of researchers working at the interface between mathematics and other scientific disciplines.

More information about this series at http://www.springer.com/series/13386

Didier Aussel · C. S. Lalitha
Editors

Generalized Nash Equilibrium Problems, Bilevel Programming and MPEC

 Springer

Editors
Didier Aussel
Department of Mathematics and Computer
  Science
University of Perpignan
Perpignan
France

C. S. Lalitha
Department of Mathematics
University of Delhi South Campus (UDSC)
New Delhi
India

ISSN 2364-6748   ISSN 2364-6756   (electronic)
Forum for Interdisciplinary Mathematics
ISBN 978-981-13-3842-7   ISBN 978-981-10-4774-9   (eBook)
https://doi.org/10.1007/978-981-10-4774-9

Printed on acid-free paper

This Springer imprint is published by the registered company Springer Nature Singapore Pte Ltd. part of
Springer Nature
The registered company address is: 152 Beach Road, #21-01/04 Gateway East, Singapore 189721,
Singapore

Preface

The book is based on the lectures delivered by experts at the International Center for Pure and Applied Mathematics (CIMPA) school on generalized Nash equilibrium problems, bilevel programming and MPEC held during November 25 to December 6, 2013, at the University of Delhi, Delhi, India. As the title suggests, this school dealt with three frontiers in applied mathematics pertaining to variational techniques and applications to economy. The objective of the school was to bring together an international team of experts working in these prominent areas of optimization. The word equilibrium is commonly used in reference to certain equilibrium phenomena that arise in engineering and economics applications. The study of equilibrium problems has been the theme of this school. Apart from generalized Nash equilibrium problems and mathematical programs with equilibrium constraints (MPECs), the school also dealt with bilevel programs which are closely related to MPECs. Bilevel programs are also known as mathematical programs with optimization constraints. Hence, it is believed that the class of MPEC is an extension of the class of bilevel programs. However, there have been studies to justify that this is not always true. When dealing with noncooperative games, the classical concept of the solution is the Nash equilibrium, a solution concept introduced by John Nash in 1950. Game problem where the strategy sets depend on the rival's strategies is termed as generalized Nash equilibrium problem (GNEP) and has applications in many fields like economics, pollution models, competitive network and wireless communication. Bilevel problem is an optimization problem whose feasible region is the solution set of another optimization problem, called lower-level problem. Bilevel problems were first formulated by H. V. Stackelberg in a monograph on market economy in 1934. MPEC is a class of constrained optimization problems where the constraints include parametric variational inequalities or complementarity problems. MPEC is also closely related to the economic problem of Stackelberg game and involves two classes of players, the leader and the follower. Bilevel programs and MPECs have many applications in the fields of economics, ecology, engineering, transportation and others.

The first three chapters of this book focus on theoretical aspects of bilevel optimization problems and mathematical programming with equilibrium problems. The first chapter is based on the lectures by S. Dempe on "Bilevel Optimization: Reformulation and First Optimality Conditions". Bilevel optimization problems considered are nonsmooth, nonconvex optimization problems whose feasible set is described by using the graph of the solution set mapping of a second parametric optimization problem. Different approaches are known to transform this problem into a one-level optimization problem. Two of the approaches are presented in this chapter. The first is based on the transformation by using the Karush–Kuhn–Tucker conditions of the convex lower-level problem resulting in an MPEC, and the second is based on the use of the optimal value function which leads to a nonsmooth optimization problem. The resulting necessary optimality conditions are derived, and first solution algorithms for the bilevel problem by using these transformations are also presented.

The second chapter is part of the lectures by R. Henrion on "Calmness as a Constraint Qualification for M-Stationarity Conditions in MPECs". Since the constraint set of MPEC is defined as the solution of some parameter-dependent generalized equations, the standard constraint qualifications of nonlinear programs are not applicable to MPECs. In this chapter, a constraint qualification to MPECs is derived as an application of calmness concept of multifunctions. In this regard, the author presents a variational analytic approach to dual stationarity conditions for MPECs on the basis of Lipschitzian properties of the perturbed generalized equation. The study also results in ensuring an appropriate calculus rule for the Mordukhovich normal cone.

The third chapter, entitled "Optimality Conditions for Bilevel Programming: An Approach Through Variational Analysis" by J. Dutta, considers the special case of a bilevel optimization problem in which the constraint set of the lower-level problem, known as the problem of the follower in economy, is described as a general closed convex set, and not necessarily by equalities and inequalities. Using recently developed tools for nonsmooth analysis, first- and second-order optimality conditions are derived.

The fourth chapter, jointly authored by B. Heymann and A. Jofré, entitled "Mechanism Design and Auctions for Electricity Networks" summarizes the results recently obtained on the modelization and analysis of deregulated electricity markets with a focus on auctions and mechanism design. It completes the series of lectures on the use of Nash games in electricity markets delivered by the second author in the research school. A special binodal bid mechanism is described in detail to illustrate the developed general analysis.

The fifth chapter by J. M. Borwein and M. K. Tam is based on a series of seven lectures titled "Techniques of variational analysis" given by the first author at the CIMPA school. The first six lectures are available online at the link http://www. carma.newcastle.edu.au/jon/ToVA/links.html. These lectures were aimed at introducing variational analysis, variational principles, nonsmooth analysis and multifunction analysis. The present chapter entitled "Reflection Methods for Inverse Problems with Applications to Protein Conformation Determination" is based on the discussions in the seventh lecture and focuses on reflection methods for protein conformation determination within the framework of matrix completion. The Douglas–Rachford reflection method is an algorithm used for solving the feasibility problem of finding a point in the intersection of finitely many sets. This chapter focuses on the application of this method to protein conformation determination where problem-specific heuristics are proposed. In the end, a complementary application of this approach in the context of ionic liquid chemistry is also demonstrated.

Finally, the sixth and last chapter by Didier Aussel on "On Single-Valuedness of Quasimonotone Set-Valued Operators" concentrates on the fundamental question of the single-valuedness of set-valued maps. Many such results for monotone set-valued have been proved in the literature. The aim of this chapter is to show that first corresponding results can be obtained for a broader class of quasimonotone maps. The concept of single-directionality plays a fundamental role in the used approach. Following the existing literature for monotone maps, three different kinds of single-directional properties are proved: pointwise, local and dense points of view. For each of them, the author has a special focus on the case of the normal operator of lower semi-continuous quasiconvex functions.

Acknowledgements We wish to thank all the authors for their contributions and all the referees whose hard work was indispensable for us to maintain the scientific quality of the volume. Special thanks go to CIMPA/ICPAM and the University of Delhi for the financial and organizational support to the research school which has contributed greatly to the success of the meeting and its outcome in the form of the present volume.

Perpignan, France   Didier Aussel
New Delhi, India   C. S. Lalitha
March 2016

Homage

In July 2016, our colleague and friend Jonathan Borwein, Professor in the University of Newcastle, Australia, passed away. We are extremely saddened and shocked by the untimely demise of such a great mathematician. His demise is an irreparable loss to the mathematics community.

Professor Borwein was one of the most renowned experts in two distinct areas, namely variational analysis and number theory. He had made many significant contributions in both these areas. He wrote several books that are now considered as classic books and would inspire generations of students and researchers.

During the CIMPA school in Delhi, J. Borwein delivered a remarkable course on variational analysis. He was an excellent teacher and easily adapted his talks to the audience, who were mainly Ph.D. students and young researchers. Equally impressive was his constant approachability. It has been always a pleasure to interact and learn from him.

Didier Aussel
C. S. Lalitha

Contents

About the Editors

Didier Aussel is a Professor in the Department of Mathematics and Computer Science, University of Perpignan, France. He is an expert on the theoretical aspects of quasiconvex optimization and variational inequalities. In addition, his research also involves applications of optimization in engineering processes and mathematical economics, in particular for the modeling of electricity markets, a topic where Nash equilibrium and multileader–follower games play a central role. He has published over 50 research articles in several prominent mathematics journals such as *Transactions of the American Mathematical Society, SIAM Journal on Control and Optimization, Journal of Optimization Theory and Applications*, as well as physics journals including *Energy Conversion and Management*. He is an Associate Editor of the journal *Optimization* and has served nearly a decade as the Co-Director and subsequently as Director of the French CNRS Research Group on Mathematics of Optimization and Applications. He has supervised several Ph.D. students, mainly on topics concerning nonsmooth variational analysis and electricity markets. Deeply interested in conveying research knowledge to young generations, he has been actively involved in the organization of research schools and research courses all over the world, including countries such as Vietnam, India, Chile, Peru, Cuba, Taiwan and Saudi Arabia.

C. S. Lalitha is a Professor in the Department of Mathematics, University of Delhi South Campus, New Delhi, India. Her areas of interest include optimization theory, nonsmooth analysis and variational inequalities. She has co-authored more than 50 research papers published in prominent journals such as *Journal of Optimization Theory and Applications, Optimization, Optimization Letters, Journal of Global Optimization* and *Journal of Mathematical Analysis and Applications*. She has also co-authored a book entitled *Generalized Convexity, Nonsmooth Inequalities and Nonsmooth Optimization* and has co-edited a book *Combinatorial Optimization: Some Aspects*, published by Narosa. She is a recipient of the INSA Teacher Award

2016. She has supervised many M.Phil. and Ph.D. students at the University of Delhi and is a member of various learned scientific societies such as the American Mathematical Society, the Operational Research Society of India, the Indian Mathematical Society and the Ramanujan Mathematical Society. She has organized many training programs, seminars and conferences. In addition, she has presented papers and delivered talks at several national and international conferences and workshops.

Chapter 1
Bilevel Optimization: Reformulation and First Optimality Conditions

Stephan Dempe

Abstract Bilevel optimization problems are nonsmooth, nonconvex optimization problems the feasible set of which is (in part) described using the graph of the solution set mapping of a second parametric optimization problem. To investigate them, their transformation into a one-level optimization problem is necessary. For that, different approaches can be used. Two of them are considered in this article: the transformation using the Karush–Kuhn–Tucker conditions of the (convex) lower level problem resulting in a mathematical program with equilibrium constraint (MPEC) and the use of the optimal value function of this problem which leads to a nonsmooth optimization problem. Besides the resulting necessary optimality conditions, first solution algorithms for the bilevel problem using these transformations are presented.

1.1 Introduction

In many practical situations, the decision makers act according to a certain hierarchical order. In the simplest situation, one decision maker, called the *leader*, has the first choice, and the other one, called the *follower*, reacts optimally to the leader's selection. Knowing the follower's reactions to his selections, it is now the leader's task to find a possible decision such that this together with the follower's reaction is a best possible one. For modeling such situations, consider the follower's problem first. This can be formulated as a parametric optimization problem:

$$\min_{y}\{f(x, y) : g(x, y) \leq 0\}, \tag{1.1}$$

where $f : \mathbb{R}^n \times \mathbb{R}^m \longrightarrow \mathbb{R}$, $g : \mathbb{R}^n \times \mathbb{R}^m \longrightarrow \mathbb{R}^p$ are sufficiently smooth. Here, x denotes the leader's selection and the follower minimizes his objective function $f(x, y)$ with respect to y subject to certain constraints $g_i(x, y) \leq 0$, $i = 1, \ldots, p$.

S. Dempe (✉)
TU Bergakademie Freiberg, Freiberg, Germany
e-mail: dempe@tu-freiberg.de

© Springer Nature Singapore Pte Ltd. 2017
D. Aussel and C. S. Lalitha (eds.), *Generalized Nash Equilibrium Problems, Bilevel Programming and MPEC*, Forum for Interdisciplinary Mathematics, https://doi.org/10.1007/978-981-10-4774-9_1

Let

$$\varphi(x) := \min_y \{f(x, y) : g(x, y) \leq 0\} : \mathbb{R}^n \longrightarrow \mathbb{R} \tag{1.2}$$

denote the *optimal value function* of the follower's problem and

$$\Psi(x) := \{y : g(x, y) \leq 0, f(x, y) \leq \varphi(x)\} : \mathbb{R}^n \rightrightarrows \mathbb{R}^m \tag{1.3}$$

the respective *solution set mapping*. $\Psi(\cdot)$ is a point-to-set mapping, mapping the leader's selection to the set of optimal solutions of the follower's problem. The follower's selection is in general not uniquely determined.

The graph of the solution set mapping is

$$\mathbf{gph}\ \Psi := \{(x, y) : y \in \Psi(x)\}.$$

Using it, the leader's problem can be formulated as

$$\min_{x,y} \{F(x, y) : G(x) \leq 0, (x, y) \in \mathbf{gph}\ \Psi\}, \tag{1.4}$$

where $F : \mathbb{R}^n \times \mathbb{R}^m \longrightarrow \mathbb{R}, G : \mathbb{R}^n \longrightarrow \mathbb{R}^q$ are supposed to be sufficiently smooth.

The origin of bilevel optimization goes back to the *Stackelberg game* (see Stackelberg [47]), its investigation in the fields of mathematical optimization started with the work of Bracken and McGill [6]. For more information about the bilevel optimization problem (1.4), the interested reader is referred to the monographs by Bard [2], Dempe [10], Dempe et al. [16], and the annotated bibliography [11] by Dempe among others.

Note that we used the so-called *optimistic formulation* of the bilevel optimization problem in (1.4) which distinguishes from the initial formulation of the problem in the sense that the leader is allowed to select a solution $y \in \Psi(x)$ out of the follower's set of optimal solutions which is a best one according to his objective function $F(\cdot, \cdot)$. The pessimistic (strong) and optimistic (weak) formulations are topic of research, e.g., by Loridan and Morgan in [32].

To investigate bilevel optimization problems, especially for deriving optimality conditions and for solving them, the problems need to be transformed into one-level optimization problems. For this, at least three approaches are possible:

1. the lower level problem (1.1) is replaced with its Karush–Kuhn–Tucker conditions,
2. the objective function of the lower level problem is bounded from above by the optimal value function, or
3. the lower level problem is transformed using a generalized equation (or a variational inequality).

Aim of the article is to investigate the first two approaches, to formulate their basic properties in relation with problem (1.4), (1.1), to present examples of corresponding necessary optimality conditions and first attempts to solve the transformed optimization problems.

Definition 1.1 A point-to-set mapping $Z : \mathbb{R}^n \rightrightarrows \mathbb{R}^m$ (mapping points $x \in \mathbb{R}^n$ to subsets in $\mathbb{R}^m$) is called upper semicontinuous at a point $x^0 \in \mathbb{R}^n$ if, for each open set $V \supset Z(x^0)$ there is an open set $U \ni x^0$ with $Z(x) \subset V$ for all $x \in U$.

If the solution set mapping $\Psi(\cdot)$ is upper semicontinuous, its graph **gph** Ψ is closed. Then, if the set $M := \{(x, y) : g(x, y) \le 0, \ G(x) \le 0\}$ is not empty and bounded, the feasible set of the bilevel optimization problem (1.4) is not empty, too, and compact. Hence, using the famous Theorem by Weierstraß we obtain

Theorem 1.1 *If the solution set mapping $\Psi(\cdot)$ of the lower level problem is upper semicontinuous and the set M is not empty and compact then problem (1.4) has an optimal solution.*

Theorem 1.2 (Bank et al. [1]) *Let $\bar{x}$ with $G(\bar{x}) \le 0$ be fixed. If for problem (1.1) the set M is not empty and compact and the Mangasarian–Fromovitz constraint qualification*

(MFCQ) Consider an optimization problem

$$\min\{f_0(x) : f_i(x) \le 0, \ i = 1, \ldots, p, \ f_i(x) = 0, \ i = p + 1, \ldots, p + q\}. \quad (1.5)$$

The Mangasarian–Fromovitz constraint qualification is satisfied for problem (1.5) at some feasible point $\hat{x}$ if

$$\exists d : \ \nabla f_i(\hat{x}) d < 0 \ \ \forall i \in I(\hat{x}) := \{j \in \{1, \ldots, p\} : f_j(\hat{x}) = 0\},$$
$$\nabla f_i(\hat{x}) d = 0 \ \ \forall i = p + 1, \ldots, p + q$$

and the gradients $\nabla f_i(\hat{x})$, $i = p + 1, \ldots, p + q$, are linearly independent.

*is satisfied at each point $(\bar{x}, \bar{y}) \in$ **gph** Ψ then, the mappings $\Psi(\cdot)$ and $\Lambda(\cdot, \cdot)$ are upper semicontinuous at $(\bar{x}, \bar{y})$ and the function $\varphi(\cdot)$ is continuous at $\bar{x}$.*

Here, $L(x, y, \lambda) = f(x, y) + \lambda^\top g(x, y)$ is the Lagrange function and

$$\Lambda(\bar{x}, \bar{y}) := \{\lambda : \ \nabla_y L(x, y, \lambda) = 0, \ \lambda \ge 0, \ \lambda^\top g(x, y) = 0\}$$

is the set of regular Lagrange multipliers of problem (1.4).

1.2 Strongly Stable Lower Level Solution

The simplest idea to investigate the bilevel optimization problem is to assume that the lower level problem (1.1) has a unique (global) optimal solution $y(x) \in \Psi(x)$ for all values of the parameter x. Assume that problem (1.1) is a convex optimization problem (for the case of nonconvex problems see Jongen and Weber [26]). Then, problem (1.4) is replaced by

$$\min_{x}\{F(x, y(x)) : G(x) \leq 0\}. \tag{1.6}$$

If the function $y(\cdot)$ admits (weak) differentiability properties, the way is paved to formulate necessary (and sufficient) optimality conditions and to describe a solution algorithm.

Definition 1.2 An optimal solution y^* of problem (1.1) is called *strongly stable in the sense of Kojima* [30] at the parameter value x^*, provided there is an open neighborhood V of x^* such that problem (1.1) has a unique optimal solution $y(x)$ for each $x \in V$ with $y(x^*) = y^*$ which is continuous at x^*.

The notion of strong stability here is formulated under the assumption of convexity. Otherwise, local optimal solutions in a fixed open set need to be investigated. To prove strong stability of the optimal solution of (1.1), some sufficient optimality condition of second order is necessary. Using this notion, the transformation of (1.4) into (1.6) is possible.

Uniqueness of optimal solution is usually verified under sufficient optimality conditions of second order. In case of parametric optimization problems, we need a strong version of such conditions:

Definition 1.3 The *strong sufficient optimality condition of second order* (SSOSC) is satisfied at some point $(\bar{x}, \bar{y})$ with $g(\bar{x}, \bar{y}) \leq 0$ if for each direction $d \neq 0$ with $\nabla g_i(\bar{x}, \bar{y})d = 0$ for each $i : \lambda_i > 0$ we have

$$d^\top \nabla^2_{yy} L(\bar{x}, \bar{y}, \lambda) d > 0.$$

In this assumption, positive definiteness of the Hessian of the Lagrange function over a subspace is used and not only that with respect to the cone of critical directions as in the usual sufficient optimality condition of second order.

The next assumption is used to guarantee that the optimal solution function $y(\cdot)$ is at least directionally differentiable.

Definition 1.4 The *constant rank constraint qualification* (CRCQ) is satisfied at the point $(\bar{x}, \bar{y})$ for the problem (1.1) if there exists an open neighborhood U of $(\bar{x}, \bar{y})$ such that, for each subset $I \subseteq \{i : g_i(\bar{x}, \bar{y}) = 0\}$ the family of gradient vectors $\{\nabla_y g_i(x, y) : i \in I\}$ has constant rank on U.

The constant rank constraint qualification guarantees existence of a regular Lagrange multiplier. It is not implied by the Mangasarian–Fromovitz constraint qualification (MFCQ) and does not imply this condition; see Minchenko and Stakhovski [35].

Theorem 1.3 (Ralph and Dempe [42]) *Assume convexity, (MFCQ), (CRCQ), and strong sufficient optimality of second order (SSOSC) for the lower level problem* (1.1).

Then, the optimal solution of the lower level problem (1.1) *is locally unique and a PC^1-function.*

Definition 1.5 A function $z : \mathbb{R}^n \to \mathbb{R}^m$ is a PC^1-function (locally around a point $\bar{x} \in \mathbb{R}^n$) if there exist an (open) neighborhood V of $\bar{x}$ and a finite number of continuously differentiable functions $z^k : V \to \mathbb{R}^m$, $k = 1, \ldots, p$:

$$z(x) \in \{z^1(x), \ldots, z^p(x)\} \ \forall \, x \in V$$

and the function z itself is continuous on V.

Example 1.1 (Dempe [10]) Consider the problem

$$-y \to \min_{y}$$
$$y \leq 1,$$
$$y^2 \leq 3 - x_1^2 - x_2^2,$$
$$(y - 1.5)^2 \geq 0.75 - (x_1 - 0.5)^2 - (x_2 - 0.5)^2,$$

with two parameters x_1 and x_2. Then, y is a continuous selection of three continuously differentiable functions $y^1 = y^1(x)$, $y^2 = y^2(x)$, $y^3 = y^3(x)$ in an open neighborhood of the point $x^0 = (1, 1)^\top$:

$$y(x) = \begin{cases} y^1 = 1, & x \in \textbf{Supp } (y, y^1), \\ y^2 = \sqrt{3 - x_1^2 - x_2^2}, & x \in \textbf{Supp } (y, y^2), \\ y^3 = 1.5 - \sqrt{0.75 - (x_1 - 0.5)^2 - (x_2 - 0.5)^2}, & x \in \textbf{Supp } (y, y^3), \end{cases}$$

where

$$\textbf{Supp } (y, y^1) = \{x : x_1^2 + x_2^2 \leq 2, \ (x_1 - 0.5)^2 + (x_2 - 0.5)^2 \geq 0.5\},$$
$$\textbf{Supp } (y, y^2) = \{x : 2 \leq x_1^2 + x_2^2 \leq 3\},$$
$$\textbf{Supp } (y, y^3) = \{x : (x_1 - 0.5)^2 + (x_2 - 0.5)^2 \leq 0.5\}.$$

PC^1-functions are locally Lipschitz continuous and directionally differentiable; see Scholtes [46]. Here, the directional derivative of a function $z : \mathbb{R}^n \to \mathbb{R}$ at $\bar{x} \in \mathbb{R}^n$ is defined as

$$z'(\bar{x}; d) := \lim_{t \downarrow 0} t^{-1}[z(\bar{x} + td) - z(\bar{x})].$$

Since $y : \mathbb{R}^m \to \mathbb{R}^n$, this definition is understood componentwise. If the (CRCQ) is violated but the other assumptions in Theorem 1.3 are satisfied, the function $y(\cdot)$ remains to be directionally differentiable but the function $d \mapsto y'(\bar{x}; d)$ is in general no longer (Lipschitz) continuous and also local Lipschitz continuity of the function $y(\cdot)$ itself is no longer guaranteed.

Let $\mathscr{F}(x) := F(x, y(x))$. If the functions F, G are continuously differentiable and the assumptions of Theorem 1.3 are satisfied, problem (1.6) is a Lipschitz optimization problem and the respective necessary optimality conditions can be used. Using Rademacher's theorem, Lipschitz continuous functions $z : \mathbb{R}^n \to \mathbb{R}$ are almost everywhere differentiable and the generalized gradient in the sense of Clarke [7] equals

$$\partial^{Cl} z(\bar{x}) = \mathbf{conv}\ \{w : \exists \{x^k\}_{k=1}^{\infty} \subset \Omega \text{ converging to } \bar{x} \text{ with } w = \lim_{k \to \infty} \nabla z(x^k)\},$$

where Ω equals the set of all points where the function z is differentiable.

Let $Z = \{x : G(x) \leq 0\}$ and

$$N_Z(\bar{x}) = \{d : \exists \lambda_i \geq 0,\ i = 1, \ldots, q,\ \lambda^{\top} G(\bar{x}) = 0,\ d = \sum_{i=1}^{q} \lambda_i \nabla G_i(\bar{x})\}$$

equals the normal cone to the set Z at $\bar{x}$, provided (MFCQ) is satisfied for problem (1.6) at this point (see, e.g., Theorem 5.2.2. in Bector et al. [4]). Moreover, if $\bar{x}$ is a local optimal solution of problem (1.6), the assumptions of Theorem 1.3 and (MFCQ) for this problem (1.6) are satisfied at $\bar{x}$, then there exist $\lambda_i,\ i = 1, \ldots, q$, such that

$$0 \in \partial^{Cl} \mathscr{F}(\bar{x}) + N_Z(\bar{x}),\ \lambda^{\top} G(\bar{x}) = 0,$$

see, for instance, Theorem 5.3.3. in Bector et al. [4].

It remains to calculate the generalized gradient of the solution function $y(\cdot)$. For that, let $y^i(\cdot),\ i = 1, \ldots, p$, denote the selection functions of the PC^1-function $y(\cdot)$ and let $\mathbf{Supp}\ (y, y^i) = \{x : y(x) = y^i(x)\}$ denote the set of all points, where the function $y(\cdot)$ equals one of its continuously differentiable selection functions.

Theorem 1.4 (Kummer [31], Scholtes [46]) *Let $z : \mathbb{R}^p \to \mathbb{R}$ be a PC^1-function. Then,*

$$\partial z(\bar{x}) = \mathbf{conv}\ \{\nabla z^i(\bar{x}) : \bar{x} \in \mathbf{cl\ int\ Supp}\ (z, z^i)\}.$$

$$I_z^0(\bar{x}) = \{i : \bar{x} \in \mathbf{cl\ int\ Supp}\ (z, z^i)\}. \tag{1.7}$$

Here $\mathbf{cl\ int\ Supp}\ (z, z^i) = \mathbf{cl\ Supp}\ (z, z^i)$, provided $\mathbf{int\ Supp}\ (z, z^i) \neq \emptyset$. The generalized derivative $\partial^{Cl} \mathscr{F}(\bar{x})$ is obtained applying a chain rule; see Clarke [7].

A bundle (trust region) algorithm can now be used to solve problem (1.6); see Outrata et al. [40]. This algorithm can be shown to converge to a Clarke stationary solution under suitable assumptions. A closely related bundle algorithm which is applied to a regularized bilevel optimization problem where the lower level problem is not assumed to have a strongly stable optimal solution can be found in the article [9] by Dempe.

A second approach to derive necessary optimality conditions for problem (1.6) uses the directional derivative itself.

Theorem 1.5 (Dempe [8]) *Let $(\bar{x}, \bar{y})$ be a local optimal solution of (1.4); the lower level problem (1.1) is assumed to be convex satisfying (MFCQ), (SSOSC), and (CRCQ) at $(\bar{x}, \bar{y})$. Then the following optimization problem has a nonnegative optimal objective function value:*

$$\alpha \rightarrow \min_{\alpha, r}$$
$$\nabla_x F(\bar{x}, \bar{y})r + \nabla_y F(\bar{x}, \bar{y})y'(\bar{x}; r) \leq \alpha$$
$$\nabla G_i(\bar{x})r \leq \alpha, \forall i : G_i(\bar{x}) = 0 \tag{1.8}$$
$$\|r\| \leq 1.$$

Moreover, if (MFCQ) is satisfied for problem (1.6), problem (1.8) can be replaced by

$$\nabla_x F(\bar{x}, \bar{y})r + \nabla_y F(\bar{x}, \bar{y})y'(\bar{x}; r) \rightarrow \min_r$$
$$\nabla G_i(\bar{x})r \leq 0, \forall i : G_i(\bar{x}) = 0 \tag{1.9}$$
$$\|r\| \leq 1.$$

The results in Theorem 1.5 can be modified to obtain a sufficient optimality condition for problem (1.4):

Theorem 1.6 (Dempe [8]) *Let $(\bar{x}, \bar{y})$ be feasible for (1.4) and assume that the other assumptions of Theorem 1.5 are satisfied.*
If the optimal function value v_1 of the problem

$$\nabla_x F(\bar{x}, \bar{y})r + \nabla_y F(\bar{x}, \bar{y})y'(\bar{x}; r) \rightarrow \min$$
$$\nabla G_i(\bar{x})r \leq 0, \forall i : G_i(\bar{x}) = 0, \tag{1.10}$$
$$\|r\| = 1$$

is strongly greater than zero $v_1 > 0$, then $(\bar{x}, \bar{y})$ is a strict local optimal solution of the bilevel problem (1.4); i.e., for arbitrary $c \in (0, v_1)$ there is $\varepsilon > 0$ such that

$$F(x, y(x)) \geq F(\bar{x}, \bar{y}) + c\|x - \bar{x}\|$$

for all x satisfying $\|x - \bar{x}\| \leq \varepsilon$, $G(x) \leq 0$.

The necessary and sufficient optimality conditions in Theorems 1.5 and 1.6 are bilevel optimization problems, too, as the following ideas for computing the directional derivative of the optimal solution $y(\cdot)$ show.

Theorem 1.7 (Ralph and Dempe [42]) *Consider problem (1.1) at a point $x = \bar{x}$ and let $\bar{y}$ be a local optimal solution of this problem where the assumptions (MFCQ), (SSOSC), and (CRCQ) are satisfied. Then the directional derivative of the function $y(\cdot)$ at $\bar{x}$ in direction r coincides with the unique optimal solution of the convex quadratic optimization problem $QP(\lambda^0, r)$*

$$0.5d^\top \nabla_{yy}^2 L(\bar{x}, \bar{y}, \lambda^0)d + d^\top \nabla_{xy}^2 L(\bar{x}, \bar{y}, \lambda^0)r \to \min_d$$

$$\nabla_x g_i(\bar{x}, \bar{y})r + \nabla_y g_i(\bar{x}, \bar{y})d \begin{cases} = 0, & \text{if } \lambda_i^0 > 0, \\ \leq 0, & \text{if } g_i(\bar{x}, \bar{y}) = \lambda_i^0 = 0 \end{cases} \quad (1.11)$$

for an arbitrary Lagrange multiplier $\lambda^0 \in \Lambda(\bar{x}, \bar{y})$ *solving*

$$\nabla_x L(\bar{x}, \bar{y}, \lambda)r \to \max_{\lambda \in \Lambda(\bar{x}, \bar{y})} . \quad (1.12)$$

A method of feasible directions in the version of Topis and Veinott [48] can be applied to solve the problem (1.6). This method is essentially based on using the necessary optimality conditions in Theorem 1.5.

Algorithm for solving problem (1.6):

Step 1 Select x^0 satisfying $G(x^0) \leq 0$, set $k := 0$, and choose $\varepsilon, \delta \in (0, 1)$.

Step 2 Compute a direction r^k, $\|r^k\| \leq 1$, satisfying

$$\mathscr{F}'(x^k; r^k) \leq s^k, \nabla G_i(x^k)r^k \leq -G_i(x^k) + s^k, \ i = 1, \ldots, q,$$

and $s^k < 0$.

Step 3 Choose a step-size t_k such that

$$\mathscr{F}(x^k + t_k r^k) \leq \mathscr{F}(x^k) + \varepsilon t_k s^k, G(x^k + t_k r^k) \leq 0.$$

Step 4 Set $x^{k+1} := x^k + t_k r^k$, compute the optimal solution $y^{k+1} = y(x^{k+1}) \in \Psi(x^{k+1})$, and set $k := k + 1$.

Step 5 If a stopping criterion is satisfied stop, else goto step 2.

The convergence result in the next theorem needs an additional nondegeneracy assumption (FRR):

(**FRR**) For each vertex $\lambda^0 \in \Lambda(x, y)$ the matrix

$$\mathscr{M} := \begin{pmatrix} \nabla_{yy}^2 L(x, y, \lambda^0) & \nabla_y^\top g_{J(\lambda^0)}(x, y) & \nabla_{xy}^2 L(x, y, \lambda^0) \\ \nabla_y g_{I_0}(x, y) & 0 & \nabla_x g_{I_0}(x, y) \end{pmatrix}$$

has full row rank $m + |I_0|$.

Here $I_0 = \{i : g_i(x, y) = 0\}$ and $J(\lambda) = \{i : \lambda_i > 0\}$.

Theorem 1.8 (Dempe and Schmidt [19]) *Consider problem (1.6), where $y(x)$ is an optimal solution of the lower level problem (1.1) which is a convex parametric optimization problem. Assume that all functions F, f, G, g are sufficiently smooth and the set $\{(x, y) : G(x) \leq 0, \ g(x, y) \leq 0\}$ is nonempty and bounded. Let the assumptions (FRR), (MFCQ), (CRCQ), and (SSOSC) for all (x, y), $y \in \Psi(x)$, $G(x) \leq 0$ be satisfied for problem (1.1) and (MFCQ) is satisfied for problem (1.6) for all x. Then, each accumulation point $(\bar{x}, \bar{y})$ of the sequence of iterates computed with the above algorithm is Clarke stationary.*

The algorithm realizes an algorithm of feasible directions in the version of Topkis and Veinott [48] and uses an Armijo step-size rule; see, e.g., Bazaraa et al. [3]. Finiteness of the number of selection functions of the lower level optimal PC^1-function $y(\cdot)$ implies that one of the selection functions is used infinitely often. If only this function is considered, the problem reduces to a differentiable optimization problem. A careful look on the optimality conditions for this problem shows that they are also necessary for a Clarke stationary solution of the problem (1.6). This shows convergence of the algorithm to a Clarke stationary solution.

1.3 Use of the Karush–Kuhn–Tucker Conditions

Let the lower level problem (1.1) be a convex one for each fixed value of the parameter x; i.e., let the functions $y \longrightarrow f(x, y)$ and $y \longrightarrow g_i(x, y)$ for all i be convex functions. Then, problem (1.4) can be replaced by the *KKT reformulation*

$$
\begin{aligned}
&F(x, y) \to \min_{x,y,u} \\
&G(x) \leq 0, \\
&\nabla_y L(x, y, \lambda) = 0, \\
&g(x, y) \leq 0, \ \lambda \geq 0, \ \lambda^\top g(x, y) = 0.
\end{aligned}
\tag{1.13}
$$

An example by Mirrlees [36] shows that this transformation is not possible if the lower level problem is not a convex optimization problem. In this example the lower level problem does not have constraints, and the global optimum of the bilevel problem is not a stationary solution of the resulting problem (1.13). It is an easy task to verify that a global optimal solution $(\bar{x}, \bar{y})$ of problem (1.4) corresponds to a global optimal solution $(\bar{x}, \bar{y}, \bar{\lambda})$ of problem (1.13) for all Lagrange multipliers $\bar{\lambda} \in \Lambda(\bar{x}, \bar{y})$ and vice versa.

The complementarity constraints $0 \leq \lambda \perp g(x, y) \geq 0$ of problem (1.13) imply that this problem has a certain combinatorial structure, its feasible set is nonconvex, problem (1.13) is a nonconvex optimization problem. Solution algorithms for such problems converge to stationary solutions. Stationary solutions of problem (1.13) do in general <u>not</u> correspond to stationary solutions of (1.4):

Theorem 1.9 (Dempe and Dutta [12]) *For a local optimal solution $(\bar{x}, \bar{y})$ of problem (1.4) where the Mangasarian–Fromovitz constraint qualification is satisfied for the lower level problem, the point $(\bar{x}, \bar{y}, \bar{\lambda})$ is a local optimal solution of problem (1.13) for all $\bar{\lambda} \in \Lambda(\bar{x}, \bar{y})$. If $(\bar{x}, \bar{y}, \bar{\lambda})$ is a local optimal solution of problem (1.13) for all $\bar{\lambda} \in \Lambda(\bar{x}, \bar{y})$, then $(\bar{x}, \bar{y})$ is a local optimal solution of problem (1.4), provided the Mangasarian–Fromovitz constraint qualification is satisfied for the lower level problem.*

If the constant rank constraint qualification is added to the assumptions of the last theorem, local optimality of $(\bar{x}, \bar{y}, \bar{\lambda})$ for problem (1.13) for all vertices $\bar{\lambda} \in \Lambda(\bar{x}, \bar{y})$

needs to be shown in order to validate local optimality of $(\bar{x}, \bar{y})$ for (1.4). The following simple example shows that it is not enough to consider only one local optimal solution of problem (1.13).

Example 1.2 (Dempe and Dutta [12]) Consider the linear lower level problem

$$\min_{y}\{-y : x + y \le 1, \ -x + y \le 1\} \tag{1.14}$$

and the upper level problem

$$\min\{(x - 1)^2 + (y - 1)^2 : (x, y) \in \mathbf{gph} \ \Psi\} \tag{1.15}$$

This problem has the unique optimal solution $(\bar{x}, \bar{y}) = (0.5, 0.5)$ and no local optimal solutions.

Consider the point $(x^0, y^0) = (0, 1)$. Here,

$$\Lambda(x, y) = \begin{cases} \{(1, 0)\} & \text{if } x > 0 \\ \{(0, 1)\} & \text{if } x < 0 \\ \text{conv}\{(1, 0), (0, 1)\} & \text{if } x = 0 \end{cases}$$

where conv A denotes the convex hull of the set A.

Then, the point $(0, 1, 0, 1)$ can be shown to be a local optimal solution of the KKT reformulation.

Problem (1.13) is a special case of a mathematical program with equilibrium constraints. The Mangasarian–Fromovitz constraint qualification is violated for problem (1.13) at every feasible point; see the article [44] by Scheel and Scholtes. This makes the verification of Karush–Kuhn–Tucker conditions for problem (1.13) difficult. Thus, different constraint qualifications (see, e.g., [25] by Hoheisel et al.) implying different types of stationary solutions of the problem (1.13) have been developed. Moreover, different relaxation schemes for solving the mathematical program with equilibrium constraints can be found together with their numerical comparison also in the above article by Hoheisel and colleagues.

Using Theorem 1.9, necessary optimality conditions for problem (1.13) can be used as necessary optimality conditions for the bilevel optimization problem (1.4), too. One such result can be found in the paper [20] by Dempe and Zemkoho. For that we need index sets to treat the complementarity slackness conditions in the Karush–Kuhn–Tucker conditions of the lower level problem:

1. $I_{-0}(x, y, \lambda) = \{i : g_i(x, y) < 0, \ \lambda_i = 0\}$,
2. $I_{00}(x, y, \lambda) = \{i : g_i(x, y) = 0, \ \lambda_i = 0\}$,
3. $I_{0+}(x, y, \lambda) = \{i : g_i(x, y) = 0, \ \lambda_i > 0\}$.

Theorem 1.10 (Dempe and Zemkoho [20]) *Let $(\bar{x}, \bar{y}, \bar{\lambda})$ be a local optimal solution of problem (1.13) and assume that the following condition is satisfied there:*

$$
\left.
\begin{aligned}
& \alpha^\top \nabla G(x) + \beta^\top \nabla_x g(x, y) + \nabla_x L(x, y, \lambda)\gamma = 0, \\
& \beta^\top \nabla_y g(x, y) + \nabla_y L(x, y, \lambda)\gamma = 0, \\
& \alpha \geq 0,\ \alpha^\top G(x) = 0, \\
& \nabla_y g_{I_{0+}(x,y,\lambda)}(x, y)\gamma = 0,\ \beta_{I_{-0}(x,y,\lambda)} = 0, \\
& (\beta_i > 0 \wedge \nabla_y g_i(\bar{x}, \bar{y})\gamma > 0) \vee \beta_i \nabla_y g_i(\bar{x}, \bar{y})\gamma = 0 \\
& \qquad\qquad \forall\, i \in I_{00}(\bar{x}, \bar{y}, \bar{\lambda})
\end{aligned}
\right\}
\Rightarrow
\begin{cases}
\alpha = 0 \\
\beta = 0 \\
\gamma = 0
\end{cases}
$$

$$(1.16)$$

Then, there exists (α, β, γ) with $\|(\alpha, \beta, \gamma)\| \leq r$ for some $r < \infty$ such that

$$
\begin{aligned}
& \nabla_x F(x, y) + \alpha^\top \nabla G(x) + \beta^\top \nabla_x g(x, y) + \nabla_x L(x, y, \lambda)\gamma = 0, \\
& \nabla_y F(x, y) + \beta^\top \nabla_y g(x, y) + \nabla_y L(x, y, \lambda)\gamma = 0, \\
& \alpha \geq 0,\ \alpha^\top G(x) = 0, \\
& \nabla_y g_{I_{0+}(x,y,\lambda)}(x, y)\gamma = 0,\ \beta_{I_{-0}(x,y,\lambda)} = 0, \\
& (\beta_i > 0 \wedge \nabla_y g_i(\bar{x}, \bar{y})^\top \gamma > 0) \vee \beta_i \nabla_y g_i(\bar{x}, \bar{y})^\top \gamma = 0\ \forall\, i \in I_{00}(\bar{x}, \bar{y}, \bar{\lambda}).
\end{aligned}
$$

These are the conditions for a so-called M-stationary solution. Other necessary optimality conditions for bilevel optimization problems using this approach via the Karush–Kuhn–Tucker conditions for the lower level problem can be found, e.g., in the Ph.D. theses [50] of Zemkoho and [22] by Franke.

One idea to solve problem (1.13) numerically is to relax the complementarity constraint $\lambda^\top g(x, y) = 0$ by $-\lambda^\top g(x, y) \leq \varepsilon$. This has been done for the first time in [45] by Scholtes. Modifications of this idea by a number of authors are compared in [25] by Hoheisel et al. It has been shown that the algorithms of this type converge under certain assumptions to certain stationary solutions where the type of the stationary solution depends on the used relaxation. Unfortunately, by Theorem 1.9, these stationary points need not to be related to stationary solutions of the bilevel optimization problem. To obtain convergence to a stationary solution of the bilevel optimization problem, more general relaxations can be used; see the Ph.D. thesis [33] by Mersha and also [34] by Mersha and Dempe. Here, problem (1.13) is replaced by

$$
\begin{aligned}
& F(x, y) \to \min_{x,y,u} \\
& G(x) \leq 0, \\
& \|\nabla_y L(x, y, \lambda)\| \leq \varepsilon, \\
& g(x, y) \leq 0,\ \lambda \geq 0, \\
& -\lambda^\top g(x, y) \leq \varepsilon.
\end{aligned}
$$

$$(1.17)$$

Clearly, if a global optimal solution $(x(\varepsilon), y(\varepsilon), \lambda(\varepsilon))$ of problem (1.17) for $\varepsilon > 0$ is computed for a sequence $\varepsilon \downarrow 0$ and this sequence $\{(x(\varepsilon), y(\varepsilon), \lambda(\varepsilon))\}_{\varepsilon \downarrow 0}$ converges to some point $(\bar{x}, \bar{y}, \bar{\lambda})$, the point $(\bar{x}, \bar{y})$ is feasible for the bilevel optimization problem (1.4) provided that the lower level problem is a convex one. In that case, it can be shown that the point $(\bar{x}, \bar{y}, \bar{\lambda})$ is a global optimal solution of (1.13) and, hence, $(\bar{x}, \bar{y})$ is a global optimal solution of the bilevel optimization problem. Convergence to a local optimal solution for the bilevel optimization problem has been shown in [33] and [34] under the assumption that the optimal solution of the lower level problem is strongly stable (see Kojima [30] for strong stability).

1.4 Optimal Value Function Transformation

Using the optimal value function $\varphi(\cdot)$ of the lower level problem defined in (1.2), problem (1.4) can be replaced by

$$
\begin{aligned}
F(x, y) &\to \min_{x,y} \\
G(x) &\leq 0, \\
f(x, y) &\leq \varphi(x) \\
g(x, y) &\leq 0
\end{aligned}
\tag{1.18}
$$

Problems (1.4) and (1.18) are fully equivalent (both w.r.t. local and global optima).

The function $\varphi(\cdot)$ is in general not differentiable, even under restrictive assumptions. Moreover, since $\kappa(x, y) = f(x, y) - \varphi(x) \geq 0$ for all $P = \{(x, y) : g(x, y) \leq 0\}$, each feasible solution of the problem (1.18) is a minimizer of the function $\kappa(\cdot, \cdot)$ over P which implies the existence of a singular Lagrange multiplier for problem (1.18). Hence, the nonsmooth Mangasarian-Fromovitz constraint qualification is violated for this problem; see Ye and Zhu [49] and Pilecka [41].

Definition 1.6 (*Ye and Zhu* [49]) Problem (1.18) is called *partially calm* at a local optimal solution (x^0, y^0) with $y^0 \in \Psi(x^0)$, $G(x^0) \leq 0$, $g(x^0, y^0) \leq 0$, if there exists $\kappa > 0$ and an open neighborhood U of $(x^0, y^0, 0)$ such that

$$
F(x, y) - F(x^0, y^0) + \kappa |u| \geq 0
$$

for all feasible solutions $(x, y, u) \in U$ for the problem

$$
\min_{x,y} \{F(x, y) : f(x, y) - \varphi(x) + u = 0, \ G(x) \leq 0, \ g(x, y) \leq 0\}
$$

Theorem 1.11 (Ye and Zhu [49]) *Problem (1.18) is partially calm at a local optimal solution (x^*, y^*) if and only if there is κ^* such that (x^*, y^*) solves the penalized problem*

$$
\min_{x,y} \{F(x, y) + \kappa(f(x, y) - \varphi(x)) : g(x, y) \leq 0, \ G(x) \leq 0\}
\tag{1.19}
$$

for all $\kappa \geq \kappa^*$.

This theorem enables one to apply necessary optimality conditions for Lipschitz continuous optimization problems provided that the optimal value function for the lower level problem is locally Lipschitz continuous. Fortunately, for smooth parametric optimization problems this is the case if a regularity condition is satisfied.

Theorem 1.12 (Klatte and Kummer [28], Gauvin and Dubeau [24]) *Consider the lower level problem* (1.1), *let the Mangasarian–Fromovitz constraint qualification be satisfied at all points* $(x, y) \in$ **gph** Ψ, *and assume the set P be not empty and bounded. Then, the optimal value function is locally Lipschitz continuous and its generalized gradient* $\partial^{Cl} \varphi(\bar{x})$ *in the sense of Clarke [7] satisfies:*

$$\partial^{Cl}(\bar{x}) \subseteq \mathbf{conv} \left\{ \bigcup_{y \in \Psi(\bar{x})} \bigcup_{\lambda \in \Lambda(\bar{x}, y)} \nabla_x L(\bar{x}, y, \lambda) \right\}.$$

Generalizations of this result can be found in Mordukhovich et al. [38, Corollary 4]: If the Mangasarian–Fromovitz constraint qualification is satisfied for the lower level problem at all points $(x, y) \in$ **gph** Ψ and the solution set mapping Ψ is *inner semi-continuos* at $(\bar{x}, \bar{y}) \in$ **gph** Ψ, we get the following upper estimate of the basic subdifferential in the sense of Mordukhovich of the optimal value function φ at $\bar{x}$:

$$\partial^M \varphi(\bar{x}) \subset \bigcup_{\lambda \in \Lambda(\bar{x}, \bar{y})} \nabla_x L(\bar{x}, \bar{y}, \lambda). \tag{1.20}$$

Theorem 1.13 *Let* $(\bar{x}, \bar{y})$ *be a local optimal solution of the bilevel optimization problem* (1.4); *let the set P be not empty and bounded. Assume that the Mangasarian–Fromovitz constraint qualification is satisfied with respect to the lower level constraints at all points* $(x, y) \in$ **gph** Ψ *and with respect to the upper level constraints* $G(x) \leq 0$ *at* $\bar{x}$. *Assume that the problem* (1.18) *is partially calm at* $(\bar{x}, \bar{y})$. *Then, there exist* $\mu \geq 0$, $\alpha \geq 0$ *with* $\mu^\top g(\bar{x}, y) = 0$, $\alpha^\top G(\bar{x}) = 0$, $\kappa \geq 0$ *and*

$$0 \in \nabla F(\bar{x}, \bar{y}) + \kappa \left[\nabla f(\bar{x}, \bar{y}) - \partial^{Cl} \varphi(\bar{x}) \times \{0\} \right] + \mu^\top \nabla g(\bar{x}, \bar{y}) + \alpha^\top \nabla G(\bar{x}) \times \{0\}.$$

The last theorem essentially goes back to Dempe et al. [13]; related results can be found in Zemkoho's Ph.D. thesis [50] or in the articles by Dempe et al. [17, 18].

To solve the bilevel optimization problem using this approach, its feasible set can be enlarged applying an upper approximation of the optimal value function. This has been done in the articles [37] by Mitsos et al., [29] by Kleniati and Adjiman, [21] by DeNegre and Ralphs. Dempe and Franke considered in [14] bilevel optimization problems with linear lower level problems. If only the objective function is parameterized in the linear lower level problem

$$\Psi_L(x) := \operatorname*{Argmin}_y \{x^\top y : By \leq d\}, \tag{1.21}$$

its optimal value function is piecewise affine linear and concave:

Theorem 1.14 (Beer [5]) *The function*

$$\varphi_L(\cdot) = \min_y \{x^\top y : By \leq d\}$$

is a piecewise linear concave function over $Q := \{x : |\varphi_L(x)| < \infty\}$. *Moreover, Q is a convex polyhedron.*

Using the superdifferential of the optimal value function (in the sense of convex analysis, see the monograph [43] by Rockafellar) we obtain an upper approximation

$$\varphi_L(x) \leq \xi_T(x) := \min\{x^\top y : y \in T\} \tag{1.22}$$

of the optimal value function for each finite set $T \subset \{y : By \leq d\}$ and all x. Moreover, by parametric linear optimization (see [39] by Nožička and colleagues) there exists a finite set $T = \widehat{T}$ such that equality is obtained for all x in (1.22). Finiteness of the set $\widehat{T}$ is a result of linear programming duality, finiteness of $\varphi_L(x)$ for some x, and the vertex property of optimal solutions of the dual problem.

Using $\xi_T(\cdot)$, the optimal objective function value of the problem

$$\min_{x,y}\{F(x, y) : G(x) \leq 0, \ By \leq d, \ x^\top y \leq \xi_T(x)\} \tag{1.23}$$

or

$$\min_{x,y}\{F(x, y) : G(x) \leq 0, \ By \leq d, \ x^\top y \leq x^\top z \ \forall z \in T\} \tag{1.24}$$

is never less than that of the optimal value problem

$$\min_{x,y}\{F(x, y) : G(x) \leq 0, \ By \leq d, \ x^\top y \leq \varphi_L(x)\}. \tag{1.25}$$

The following algorithm can be used to implement this idea: Using the same ideas

Algorithm solving problem (1.25):

Step 0   Fix a finite subset $T^1 \subset \{y : By \leq d\}$. Set $k := 1$.
Step 1   Solve problem (1.24) with $T = T^k$ globally. Let (x^k, y^k) denote an optimal solution.
Step 2   If (x^k, y^k) is feasible for (1.25), stop. Otherwise, compute an optimal solution z^k of the lower level problem with the parameter $x = x^k$. Set $T^{k+1} := T^k \cup \{z^k\}, k := k + 1$ and go to Step 1.

as in [14] by Dempe and Franke, it is easy to show that the above algorithm converges to a global optimal solution, provided it computes an infinite sequence of iterates and the set $M = \{(x, y) : G(x) \leq 0, \ By \leq d\}$ is not empty and bounded.

Problem (1.24) is again a nonconvex optimization problem. For such problems, it is $\mathcal{NP}$-hard (see Garey and Johnson [23] for the definition) to compute a global optimal solution in general. Standard solution algorithms compute locally optimal or even only stationary solutions. Hence, modify the above algorithm by replacing Step 1 by

Step 1': Compute a stationary solution of problem (1.24). Let (x^k, y^k) denote a stationary point.

Let the set M be not empty and bounded. Assume that the above algorithm computes an infinite sequence of stationary points. Then, by finiteness of the set $\widehat{T}$, the function $\xi_T(\cdot)$ remains unchanged after a finite number k of iterations: $\xi_{T^k}(x) = \xi_{T^q}(x)$ for all x and $q \geq k$. This implies that the feasible set M^q of problem (1.24) solved in Step 1' remains constant starting with iteration k: $M^q = M^k$ for all $q \geq k$. Now, assume that the algorithm computes an infinite sequence $\{(x^t, y^t)\}_{t=1}^{\infty}$ of stationary points converging to $(\bar{x}, \bar{y})$.

1. Then, starting with iteration k the feasible set of problem (1.24) solved in Step 1' remains unchanged. Hence, all points (x^t, y^t) with $t \geq k$ are feasible points for the same problem:
$$\min\{F(x, y) : (x, y) \in M^k\}$$

   for all $t \geq k$. This indicates that the function $\varphi_L(\cdot)$ is approximated locally by $\xi(\cdot)$ sufficiently good.
2. Let the point $(\bar{x}, \bar{y})$ satisfy a sufficient optimality condition of first order: There exist $\varepsilon > 0, \delta > 0$ such that $F(x, y) \geq F(\bar{x}, \bar{y}) + \delta\|(x, y) - (\bar{x}, \bar{y})\|$ for all feasible points (x, y) with $\|(x, y) - (\bar{x}, \bar{y})\| \leq \varepsilon$. This implies that the local optimal solution $(\bar{x}, \bar{y})$ of problem (1.25) is computed already in iteration $k + 1$.

Theorem 1.15 (Dempe and Franke [14]) *If the set M is not empty and bounded and a sufficient optimality condition of first order is satisfied at each stationary point of problem* (1.25), *then the above algorithm stops after a finite number of iterations with computing a local optimal solution.*

Consider now the optimal value problem with a right-hand side perturbed linear lower level problem having the solution set mapping

$$\Psi_R(x) = \underset{y}{\text{Argmin}}\, \{c^\top y : Ay = x, \ y \geq 0\}. \tag{1.26}$$

The optimal value function of this problem is again piecewise affine linear, but now convex, cf. Beer [5]. Let T be a convex polytope with vertices $x^1, \ldots, x^q$ and

$$\{x : G(x) \leq 0\} \subseteq T.$$

Let $\varphi_R(x) = \min_y\{c^\top y : Ay = x, \ y \geq 0\}$ denote the optimal value function of problem (1.26), and assume without loss of generality that $Q := \{x : |\varphi_R(x)| < \infty\} = T$. Each $x \in Q$ is a convex combination of the vertices of the set T and $\varphi_R(x) \leq \xi(x)$ with

$$\xi(x) = \min_{\mu} \left\{ \sum_{k=1}^{q} \mu_k \varphi_R(x^k) : \sum_{k=1}^{q} \mu_k x^k = x, \ \sum_{k=1}^{q} \mu_k = 1, \ \mu_k \geq 0, \ k = 1, \ldots, q \right\}$$
(1.27)

by convexity. The best upper estimation of the function $\varphi_R(\cdot)$ is computed solving the linear optimization problem in (1.27). Let $(\mu_B(x), \mu_N(x))$ be an optimal basic solution of this problem with the basic matrix B: $\mu_B(x) = B^{-1}x$, $\mu_N(x) = 0$. Here, B is a quadratic, regular submatrix of the matrix X composed by the vectors x^k: $X = (x^1 \ x^2 \ \ldots \ x^q)$.

Let $\mathscr{B}(X)$ denote the set of all basic matrices of X used to compute optimal basic solutions of the problem in (1.27) for $x \in T$. Denote the set of all x for which an optimal basic solution of the problem in (1.27) is obtained for one fixed basic matrix $B \in \mathscr{B}(X)$ by

$$\mathscr{R}(B) = \{x \in T : (\mu_B(x), \mu_N(x)) \text{ with } \mu_B(x) = B^{-1}x \geq 0, \ \mu_N(x) = 0$$
$$\text{solves the problem in (1.27)}\}.$$

$\mathscr{R}(B)$ is the *region of stability* for the basic matrix $B \in \mathscr{B}(X)$ and

$$T = \bigcup_{B \in \mathscr{B}(X)} \mathscr{R}(B).$$

The region of stability is a convex polytope given as the solution set of a system of linear inequalities, and $\xi(\cdot)$ is a linear function over $\mathscr{R}(B)$. Now,

$$\{(x, y) : G(x) \leq 0, \ (x, y) \in \mathbf{gph} \ \Psi_R\}$$
$$= \{(x, y) : G(x) \leq 0, \ c^\top y \leq \varphi_R(x), \ Ay = x, \ y \geq 0\}$$
$$\subseteq \{(x, y) : G(x) \leq 0, \ c^\top y \leq \xi(x), \ Ay = x, \ y \geq 0\}$$
$$= \bigcup_{B \in \mathscr{B}(X)} \{(x, y) : G(x) \leq 0, \ c^\top y \leq \xi(x), \ x \in \mathscr{R}(B), \ Ay = x, \ y \geq 0\}.$$

This implies that

$$\min_{B \in \mathscr{B}(X)} \min_{x,y} \{F(x, y) : G(x) \leq 0, \ c^\top y \leq \xi(x), \ x \in \mathscr{R}(B), \ Ay = x, \ y \geq 0\} \quad (1.28)$$

is a lower bound for the optimal objective function value of the bilevel optimization problem

$$\min_{x,y} \{F(x, y) : G(x) \leq 0, \ (x, y) \in \mathbf{gph} \ \Psi_R\}. \quad (1.29)$$

Let $(\hat{x}, \hat{y})$ be a global optimal solution of problem (1.28). Then, $\hat{x} \in \mathscr{R}(\widehat{B})$ for some $\widehat{B} \in \mathscr{B}(X)$. If $c^\top \hat{y} \leq \varphi_R(\hat{x})$, the point $(\hat{x}, \hat{y})$ is feasible for problem (1.29) and, hence, globally optimal. In the opposite case, the vector $\hat{x}$ is appended to the matrix X

resulting in matrix $\widehat{X}$. The set $\mathscr{B}(\widehat{X})$ of all basic matrices of the matrix $\widehat{X}$ is a superset of $\mathscr{B}(X) \setminus \widehat{B}$. Computing the best approximation $\xi(\cdot)$ by solving the parametric optimization problem (1.27) where the vectors x^k are now column vectors of the matrix $\widehat{X}$ results in the following.

All regions of stability $\mathscr{R}(B)$ with $B \in \mathscr{B}(X) \setminus \widehat{B}$ remain unchanged. The region $\mathscr{R}(\widehat{B})$ is partitioned into subsets whose union is equal to $\mathscr{R}(\widehat{B})$. The optimal function values of the problem (1.27) decreases over the set $\mathscr{R}(\widehat{B})$ which implies that the optimal function value of

$$\min_{x,y}\{F(x, y) : G(x) \leq 0, \ c^\top y \leq \xi(x), \ x \in \mathscr{R}(\widehat{B}), \ Ay = x, \ y \geq 0\} \qquad (1.30)$$

increases.

Summing up, this approach results in an enumeration procedure for solving the bilevel optimization problem (1.29): Note that the best bound rule is realized in the

Algorithm solving problem (1.29):

Step 0   Compute the function $\xi(\cdot)$ by solving the parametric optimization problem 1.27, compute the set $\mathscr{B}(X)$ and the regions of stability $\mathscr{R}(B)$ for all $B \in \mathscr{B}(X)$. Solve the problems (1.30) for all $B \in \mathscr{B}(X)$ and compute a solution of problem (1.28).

Step 1   Let $(\widehat{x}, \widehat{y})$ be an optimal solution of problem (1.28). If $c^\top \widehat{y} \leq \varphi_R(\widehat{x})$, the point $(\widehat{x}, \widehat{y})$ is a global optimal solution of the bilevel optimization problem (1.29).

Step 2   Otherwise append $\widehat{x}$ to X, partition the set $\mathscr{R}(\widehat{B})$, compute the function $\xi(\cdot)$ over the set $\mathscr{R}(\widehat{B})$ and goto Step 1.

above enumeration algorithm. Let the set $\{x : G(x) \leq 0\}$ be not empty and bounded. Then, due to parametric linear optimization, the function $\varphi_R(\cdot)$ is a piecewise affine-linear convex function with only a finite number of affine-linear pieces. Hence, there exist only a finite number of regions of stability which are calculated after a finite number of iterations of the above algorithm. Over each of these regions of stability, a global optimal solution of problem (1.30) is calculated and the best of these solutions is a global optimal solution of the bilevel optimization problem (1.28). This implies

Theorem 1.16 (Dempe and Franke [15]) *If the set $\{x : G(x) \leq 0\}$ is not empty and bounded, the above algorithm computes a global optimal solution of the bilevel optimization problem (1.29) and stops after a finite number of iterations.*

Related ideas can be found in the article [27] by Kalashnykova and colleagues.

1.5 Exercises

1. Solve problem (1.14), (1.15) using the different approaches described in this chapter.
2. Solve problem

$$\min_{x,y} -x - 2y$$

$$\text{subject to} \quad \begin{aligned} 2x - 3y &\geq -12 \\ x + y &\leq 14 \end{aligned}$$

$$\text{and } y \in \operatorname*{Argmin}_{y} \{-y : -3x + y \leq -3, \ 3x + y \leq 30\}.$$

again using different approaches.
3. Solve the problem in the last exercise after shifting the upper level constraints to the lower level problem.

References

1. Bank, B., Guddat, J., Klatte, D., Kummer, B., Tammer, K.: Non-linear Parametric Optimization. Birkhäuser Verlag, Basel, Boston, Stuttgart (1983)
2. Bard, J.F.: Practical bilevel optimization: algorithms and applications. Kluwer Academie Publishers, Dordrecht (1998)
3. Bazaraa, M.S., Sherali, H.D., Shetty, C.M.: Nonlinear Programming. Theory and Algorithms, vol. xv, 3rd edn. Wiley, Hoboken, NJ (2006) (English)
4. Bector, C.R., Chandra, S., Dutta, J.: Principles of Optimization Theory. Alpha Science, UK (2005)
5. Beer, K.: Lösung großer linearer Optimierungsaufgaben. Deutscher Verlag der Wissenschaften, Berlin (1977)
6. Bracken, J., McGill, J.: Mathematical programs with optimization problems in the constraints. Oper. Res. 21, 37–44 (1973)
7. Clarke, F.H.: Optimization and Nonsmooth Analysis. Wiley, New York (1983)
8. Dempe, S.: A necessary and a sufficient optimality condition for bilevel programming problems. Optimization 25, 341–354 (1992)
9. Dempe, S.: A bundle algorithm applied to bilevel programming problems with non-unique lower level solutions. Comput. Optim. Appl. 15, 145–166 (2000)
10. Dempe, S.: Foundations of Bilevel Programming. Kluwer Academie Publishers, Dordrecht (2002)
11. Dempe, S.: Annotated bibliography on bilevel programming and mathematical programs with equilibrium constraints. Optimization 52, 333–359 (2003)
12. Dempe, S., Dutta, J.: Is bilevel programming a special case of a mathematical program with complementarity constraints? Math. Program. 131, 37–48 (2012)
13. Dempe, S., Dutta, J., Mordukhovich, B.S.: New necessary optimality conditions in optimistic bilevel programming. Optimization 56, 577–604 (2007)
14. Dempe, S., Franke, S.: Solution algorithm for an optimistic linear stackelberg problem. Comput. Oper. Res. 41, 277–281 (2014)
15. Dempe, S., Franke, S.: On the solution of convex bilevel optimization problems. Comput. Optim. Appl. 63, 685–703 (2016)

16. Dempe, S., Kalashnikov, V., Pérez-Valdés, G.A., Kalashnykova, N.: Bilevel Programming Problems—Theory. Algorithms and Applications to Energy Networks. Springer, Heidelberg (2015)
17. Dempe, S., Mordukhovich, B.S., Zemkoho, A.B.: Sensitivity analysis for two-level value functions with applications to bilevel programming. SIAM J. Optim. **22**, 1309–1343 (2012)
18. Dempe, S., Mordukhovich, B.S., Zemkoho, A.B.: Necessary optimality conditions in pessimistic bilevel programming. Optimization **63**(4), 505–533 (2014)
19. Dempe, S., Schmidt, H.: On an algorithm solving two-level programming problems with nonunique lower level solutions. Comput. Opt. Appl. **6**, 227–249 (1996)
20. Dempe, S., Zemkoho, A.B.: On the Karush-Kuhn-Tucker reformulation of the bilevel optimization problem. Nonlinear Anal. Theory Methods Appl. **75**, 1202–1218 (2012)
21. DeNegre, S.T., Ralphs, T.K.: A branch-and-cut algorithm for integer bilevel linear programs. In: Chinneck, J.W., Kristjansson, B., Saltzman, M.J. (eds.) Operations Research and Cyber-Infrastructure. Operations Research/Computer Science Interfaces, vol. 47, pp. 65–78. Springer, USA (2009)
22. Franke, S.: The bilevel programming problem: optimal value and Karush-Kuhn-Tucker reformulation. Ph.D. thesis, TU Bergakademie Freiberg (2014)
23. Garey, M.R., Johnson, D.S.: Computers and Intractability: A Guide to the Theory of NP-Completeness. W.H. Freeman and Co., San Francisco (1979)
24. Gauvin, J., Dubeau, F.: Differential properties of the marginal function in mathematical programming. Math. Program. Study **19**, 101–119 (1982)
25. Hoheisel, T., Kanzow, C., Schwartz, A.: Theoretical and numerical comparison of relaxation methods for mathematical programs with complementarity constraints. Math. Program. **137**(1–2), 257–288 (2013)
26. Jongen, H.Th: Weber, G.-W: Nonlinear optimization: characterization of structural optimization. J. Glob. Optim. **1**, 47–64 (1991)
27. Kalashnykova, N.I., Kalashnikov, V.V., Dempe, S., Franco, A.A.: Application of a heuristic algorithm to mixed-integer bi-level programming problems. Int. J. Innovative Compu. Inf. Control **7**(4), 1819–1829 (2011)
28. Klatte, D., Kummer, B., Stability properties of infima and optimal solutions of parametric optimization problems. In: Demyanov, V.F. (ed.) Nondifferentiable Optimization: Motivations and Applications, Proceedings of the IIASA Workshop, Sopron, 1984. Lecture Notes in Economics and Mathematical Systems, vol. 255, pp. 215–229. Springer, Berlin (1984)
29. Kleniati, P.-M., Adjiman, C.S.: Branch-and-sandwich: a deterministic global optimization algorithm for optimistic bilevel programming problems. Part II: convergence analysis and numerical results. J. Glob. Optim. **60**(3), 459–481 (2014)
30. Kojima, M.: Strongly stable stationary solutions in nonlinear programs. In: Robinson, S.M. (ed.) Analysis and Computation of Fixed Points, pp. 93–138. Academic Press, New York (1980)
31. Kummer, B.: Newton's method for non-differentiable functions. Advances in Mathematical Optimization, Mathematical Research, vol. 45. Akademie-Verlag, Berlin (1988)
32. Loridan, P., Morgan, J.: Weak via strong Stackelberg problem: new results. J. Glob. Optim. **8**, 263–287 (1996)
33. Mersha, A.G.: Solution methods for bilevel programming problems. Ph.D. thesis, TU Bergakademie Freiberg (2008)
34. Mersha, A.G., Dempe, S.: Feasible direction method for bilevel programming problem. Optimization **61**(4–6), 597–616 (2012)
35. Minchenko, L., Stakhovski, S.: Parametric nonlinear programming problems under the relaxed constant rank condition. SIAM J. Optim. **21**(1), 314332 (2011)
36. Mirrlees, J.A.: The theory of moral hazard and unobservable behaviour: part I. Rev. Econ. Stud. **66**, 3–21 (1999)
37. Mitsos, A., Chachuat, B., Barton, P.I.: Towards global bilevel dynamic optimization. J. Glob. Optim. **45**(1), 63–93 (2009)

38. Mordukhovich, B.S., Nam, N.M., Yen, N.D.: Subgradients of marginal functions in parametric mathematical programming. Math. Program. **116**, 369–396 (2009)
39. Nožička, E., Guddat, J., Hollatz, H., Bank, B.: Theorie der linearen parametrischen Optimierung. Akademie-Verlag, Berlin (1974)
40. Outrata, J., Kočvara, M., Zowe, J.: Nonsmooth Approach to Optimization Problems with Equilibrium Constraints. Kluwer Academic Publishers, Dordrecht (1998)
41. Pilecka, M.: Combined reformulation of bilevel programming problems. Master's thesis, TU Bergakademie Freiberg, Fakultät für Mathematrik und Informatik (2011)
42. Ralph, D., Dempe, S.: Directional derivatives of the solution of a parametric nonlinear program. Math. Program. **70**, 159–172 (1995)
43. Rockafellar, R.T.: Convex Analysis. Princeton University Press, Princeton (1970)
44. Scheel, H., Scholtes, S.: Mathematical programs with equilibrium constraints: stationarity, optimality, and sensitivity. Math. Oper. Res. **25**, 1–22 (2000)
45. Scholtes, S.: Convergence properties of a regularization scheme for mathematical programs with complementarity constraints. SIAM J. Optim. **11**, 918–936 (2001)
46. Scholtes, S.: Introduction to Piecewise Differentiable Equations. Springer, New York (2012)
47. Stackelberg, H.V.: Marktform und Gleichgewicht [english translation: The Theory of the Market Economy]. Springer, Berlin (1934), Oxford University Press (1952)
48. Topkis, D.M., Veinott, A.F.: On the convergence of some feasible direction algorithms for nonlinear programming. SIAM J. Control **5**, 268–279 (1967)
49. Ye, J.J., Zhu, D.L.: Optimality conditions for bilevel programming problems. Optimization **33**, 9–27 (1995)
50. Zemkoho, A.B.: Bilevel programming: Reformulations, regularity, and stationarity. Ph.D. thesis, TU Bergakademie Freiberg (2012)

Chapter 2
Calmness as a Constraint Qualification for M-Stationarity Conditions in MPECs

René Henrion

Abstract Mathematical programs with equilibrium constraints (MPECs) represent an important class of nonlinear optimization problems. Due to their constraint set being defined as the solution of some parameter-dependent generalized equation, the application of standard constraint qualifications (CQs) from nonlinear programming to MPECs is not straightforward. Rather than turning MPECs into mathematical programs with complementarity constraints (MPCCs) and applying specially adapted CQs, we want to present here a variational-analytic approach to dual stationarity conditions for MPECs on the basis of Lipschitzian properties of the perturbed generalized equation. The focus will be on the so-called calmness property, ensuring an appropriate calculus rule for the Mordukhovich normal cone.

2.1 Introduction

This chapter is devoted to a rather self-contained introduction to the *calmness* concept of multifunctions and its application as a constraint qualification to *Mathematical Programs with Equilibrium Constraints*, MPECs for short. Here, under a constraint qualification we understand a property ensuring the derivation of dual necessary optimality conditions. We shall follow a variational-analytic approach to this problem. For this purpose, we consider an MPEC as a special case of an abstract optimization problem

$$\min\{f(x)|G(x) \in C\} \quad f : \mathbb{R}^n \to \mathbb{R}; \quad G : \mathbb{R}^n \to \mathbb{R}^p; \quad C \subseteq \mathbb{R}^p, \qquad (2.1)$$

where the objective f and the constraint mapping G are continuous and the set C is supposed to be closed. An obvious instance of (2.1) is a conventional nonlinear optimization problem with equality and inequality constraints, which results upon

R. Henrion (✉)
Weierstrass Institue for Applied Analysis and Stochstics, Berlin, Germany
e-mail: henrion@wias-berlin.de

© Springer Nature Singapore Pte Ltd. 2017
D. Aussel and C. S. Lalitha (eds.), *Generalized Nash Equilibrium Problems, Bilevel Programming and MPEC*, Forum for Interdisciplinary Mathematics, https://doi.org/10.1007/978-981-10-4774-9_2

putting $C := \{0\}_{p_1} \times \mathbb{R}_+^{p_2}$ with $p_1 + p_2 = p$ and f, G being continuously differentiable.

An MPEC is an optimization problem whose constraint is given by a parameter-dependent generalized equation:

$$\min\{\varphi(x, y) | 0 \in F(x, y) + N_\Gamma(y)\} \quad \varphi : \mathbb{R}^{n+m} \to \mathbb{R}, \; F : \mathbb{R}^{n+m} \to \mathbb{R}^m, \quad (2.2)$$

Here, $\Gamma \subseteq \mathbb{R}^m$ is closed and 'N' refers to an appropriate normal cone (e.g. normal cone of convex analysis if Γ is convex or Fréchet normal cone). Such problems have a wide range of applications in mechanics or economy (e.g. in the description of equilibria in electricity spot markets [3]). In order to derive dual stationarity conditions for (2.2), we provide first some introduction to some necessary concepts of nonsmooth calculus and to Lipschitzian properties of set-valued mappings. The role of calmness as a constraint qualification is illustrated then before discussing several options to check this property. Finally, M-stationarity conditions are derived and made fully explicit.

2.2 Some Tools from Variational Analysis

We recall that a set-valued mapping $\Phi : X \rightrightarrows Y$ between topological spaces X, Y is a conventional (single-valued) mapping $\Phi : X \to 2^Y$ assigning to each $x \in X$ a subset $\Phi(x) \subseteq Y$. A set-valued mapping is uniquely defined by its graph

$$\mathrm{gr}\, \Phi := \{(x, y) \in X \times Y \mid y \in \Phi(x)\}.$$

The inverse of Φ is defined as

$$\Phi^{-1}(y) := \{x \in X \mid y \in \Phi(x)\}.$$

2.2.1 Elements of Nondifferentiable Calculus

We recall the definition of the contingent cone and the Fréchet normal cone to a closed set:

Definition 2.1 Let $C \subseteq \mathbb{R}^n$ be closed and $\bar{x} \in C$. The **contingent cone** and the **Fréchet normal cone**, respectively, to C at $\bar{x}$ are defined as

$$T_C(\bar{x}) := \{d \in \mathbb{R}^n \mid \exists t_n \downarrow 0 \; \exists x_n \in C : t_n^{-1}(x_n - \bar{x}) \to d\}$$
$$\widehat{N}_C(\bar{x}) := \{x^* \in \mathbb{R}^n \mid \langle x^*, d \rangle \leq 0 \; \forall d \in T_C(\bar{x})\}.$$

Fig. 2.1 Illustration of the contingent and the Fréchet normal cone to a closed set C at some $\bar{x} \in C$

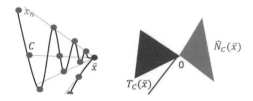

For an illustration, see Fig. 2.1. Clearly, the contingent cone is nonconvex in general, whereas the Fréchet normal cone as its negative polar cone is always convex. In order to define $T_C, \widehat{N}_C$ as set-valued mappings, we formally put $T_C(\bar{x}) := \emptyset$ and $\widehat{N}_C(\bar{x}) := \emptyset$, whenever $\bar{x} \notin C$.

Exercise 2.1 Show the following statements:

1. If C is a closed cone, then $T_C(0) = C$.
2. $T_{\mathbb{R}^n_+}(x) = \{h \in \mathbb{R}^n \mid x_i > 0 \Rightarrow h_i \geq 0 \;\; \forall i = 1, \ldots, n\}$.
3. $\widehat{N}_{\mathbb{R}^n_+}(x) = \{x^* \in \mathbb{R}^n_- \mid \langle x^*, x \rangle = 0\}$.

Since, by convention, $\widehat{N}_C(x) = \emptyset$ for $x \notin C$, 3. in Exercise 2.1 amounts to

$$\operatorname{gr} \widehat{N}_{\mathbb{R}^n_+} = \{(x, x^*) \in \mathbb{R}^n_+ \times \mathbb{R}^n_- \mid \langle x, x^* \rangle = 0\}. \tag{2.3}$$

If the closed set C happens to be convex, then the Fréchet normal cone to C coincides with the normal cone of convex analysis. The use of the Fréchet normal cone suffers from a lack of good calculus rules due to its graph being not closed in general. Therefore, it makes sense rather to consider a normal cone whose graph is the closure of the graph of the Fréchet normal cone [6]. Translating this into an explicit definition yields.

Definition 2.2 Let $C \subseteq \mathbb{R}^n$ be closed and $\bar{x} \in C$. The **Mordukhovich normal cone** to C at $\bar{x}$ is defined as

$$N_C(\bar{x}) := \{x^* \mid \exists (x_n, x_n^*) \to (\bar{x}, x^*) : x_n \in C, \; x_n^* \in \widehat{N}_C(x_n)\}$$

Figure 2.2 illustrates the computation of the Mordukhovich normal cone to some closed set C at one of its points $\bar{x}$. In a first step, Fréchet normal cones to C in a neighbourhood of $\bar{x}$ are computed. In the example, these are onedimensional linear subspaces for points $x \neq \bar{x}$ (normals to curves in the sense of classical analysis) and a solid polyhedral cone in $\bar{x}$ itself. In the second step, all limits of such Fréchet normals are aggregated according to Definition 2.2 to yield the Mordukhovich normal cone, which in contrast to the Fréchet normal cone may be nonconvex (see Fig. 2.2). Similar to the Fréchet normal cone, the Mordukhovich normal cone coincides with the one of convex analysis for convex sets.

Example 2.1 As an illustration, we compute the normal cone to different points of the set $C := \operatorname{gr} N_{\mathbb{R}_+} \subseteq \mathbb{R}^2$. By convexity of $\mathbb{R}_+$ and by (2.3), we have that

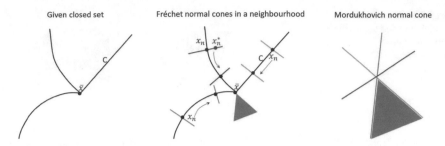

Fig. 2.2 Given closed set C and point $\bar{x} \in C$ (left), all Fréchet normal cones in a neighbourhood of $\bar{x}$ (middle) and Mordukhovich normal cone in $\bar{x}$ (right)

Fig. 2.3 Illustration of the sets $\mathrm{gr}\, N_{\mathbb{R}_+}$ (left) and $N_{\mathrm{gr}\, N_{\mathbb{R}_+}}(0,0)$ (right)

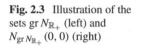

$$\mathrm{gr}\, N_{\mathbb{R}_+} = \mathrm{gr}\, \widehat{N}_{\mathbb{R}_+} = [\mathbb{R} \times \{0\}] \cup [\{0\} \times \mathbb{R}_-] \tag{2.4}$$

(see Fig. 2.3). Hence, there are three possibilities for a point $\bar{x}$ belonging to $\mathrm{gr}\, N_{\mathbb{R}_+}$: first, one may have that $\bar{x}_1 > 0$ and $\bar{x}_2 = 0$. Then, $T_{\mathrm{gr}\, N_{\mathbb{R}_+}}(\bar{x}) = \mathbb{R} \times \{0\}$ and, hence, $\widehat{N}_{\mathrm{gr}\, N_{\mathbb{R}_+}}(\bar{x}) = \{0\} \times \mathbb{R}$. Similarly, in the second case, $\bar{x}_1 = 0$ and $\bar{x}_2 > 0$, we derive that $\widehat{N}_{\mathrm{gr}\, N_{\mathbb{R}_+}}(\bar{x}) = \mathbb{R} \times \{0\}$. Finally, for the remaining third case, $\bar{x} = (0,0)$, the fact that $\mathrm{gr}\, N_{\mathbb{R}_+}$ is a closed cone implies via 1. of Exercise 2.1 that $T_{\mathrm{gr}\, N_{\mathbb{R}_+}}(0,0) = \mathrm{gr}\, N_{\mathbb{R}_+}$. Consequently,

$$\widehat{N}_{\mathrm{gr}\, N_{\mathbb{R}_+}}(0,0) = \{x^* \mid \langle x^*, h \rangle \leq 0 \quad \forall h \in \mathrm{gr}\, N_{\mathbb{R}_+}\} = \mathbb{R}_- \times \mathbb{R}_+,$$

where the last equation follows from (2.4). Now, aggregating all limits of Fréchet normals in the neighbourhood of $\bar{x}$ in the sense of Definition 2.2 amounts in our example simply to collecting the union of Fréchet normal cones in the three discussed cases. Hence, at $\bar{x} = (0,0)$ we have that (see Fig. 2.3)

$$N_{\mathrm{gr}\, N_{\mathbb{R}_+}}(0,0) = [\mathbb{R}_- \times \mathbb{R}_+] \cup [\{0\} \times \mathbb{R}] \cup [\mathbb{R} \times \{0\}].$$

As the first and second cases considered above ($\bar{x} \neq (0,0)$) are stable (i.e. remain unchanged under a small perturbation of $\bar{x} \in \mathrm{gr}\, N_{\mathbb{R}_+}$), the Fréchet normal cones are locally constant around $\bar{x}$ and, consequently, coincide with the Mordukhovich normal cone.

An important property of the normal cone is that it commutes with the Cartesian product (see [6, Proposition 1.2]):

$$N_{C_1 \times \cdots \times C_n}(\bar{x}_1, \ldots, \bar{x}_n) = N_{C_1}(\bar{x}_1) \times \cdots \times N_{C_n}(\bar{x}_n) \tag{2.5}$$

Exercise 2.2 Using the Cartesian product formula above, show that

$$\text{gr} \, N_{\mathbb{R}_+^p} = L^{-1}(\Lambda),$$

where

$$L(x_1, \ldots, x_p, y_1, \ldots, y_p) := (x_1, y_1, \ldots, x_p, y_p), \quad \Lambda := \text{gr} \, N_{\mathbb{R}_+} \times \cdots \times \text{gr} \, N_{\mathbb{R}_+}.$$

As usual in nondifferentiable calculus, a normal cone induces a subdifferential of lower semicontinuous (possibly extended-valued) functions:

Definition 2.3 Let $f : \mathbb{R}^n \to \mathbb{R} \cup \{\infty\}$ be lower semicontinuous and define (closed) epigraph as $\text{epi} f := \{(x, t) \in \mathbb{R}^{n+1} \mid t \geq f(x)\}$. Then, the **subdifferential** of f at $\bar{x}$ is defined as

$$\partial f(\bar{x}) := \{x^* \in \mathbb{R}^n | (x^*, -1) \in N_{\text{epi} f}(\bar{x}, f(\bar{x}))\}.$$

Analogous to the normal cone, the subdifferential is nonconvex in general, but coincides with the subdifferential of convex analysis for convex functions. If f happens to be continuously differentiable, then $\partial f(\bar{x}) = \nabla f(\bar{x})$.

Exercise 2.3 Show that for $f(x) := -|x|$ one has that $\partial f(0) = \{-1, 1\}$. Hint: Verify that

$$\widehat{N}_{\text{epi} f}(x, t) = \begin{cases} (0, 0) & \text{if } t > f(x) \text{ or } x = t = 0 \\ \mathbb{R}_+(1, -1)) & \text{if } t = f(x) \text{ and } x < 0 \\ \mathbb{R}_+(-1, -1)) & \text{if } t = f(x) \text{ and } x > 0 \end{cases}$$

Using this, aggregate Fréchet normals for $(x, t) \in \text{epi} f$ in a neighbourhood of $(\bar{x}, f(\bar{x}))$ in order to derive that $N_{\text{epi} f}(\bar{x}, f(\bar{x})) = \text{gr} f$. Apply Definition 2.3.

The subdifferential satisfies the following important sum rule

Theorem 2.1 (see [6], Theorem 2.33) *Let $f_1 : \mathbb{R}^n \to \mathbb{R}$ be locally Lipschitzian, and let $f_2 : \mathbb{R}^n \to \mathbb{R}$ be lower semicontinuous. Then,*

$$\partial(f_1 + f_2)(\bar{x}) \subseteq \partial f_1(\bar{x}) + \partial f_2(\bar{x}).$$

The normal cone and the subdifferential introduced so far can be employed in order to state the following necessary optimality conditions for an abstract optimization problem:

Theorem 2.2 (see [8], Theorem 8.15) *Let $f : \mathbb{R}^n \to \mathbb{R}$ be locally Lipschitzian, and let $\bar{x}$ be a local solution of the optimization problem*

$$\min\{f(x) | x \in C\} \quad (C \subseteq \mathbb{R}^n \text{closed})$$

Then, $0 \in \partial f(\bar{x}) + N_C(\bar{x})$.

We note that a similar formula would not be valid for Fréchet normal cones.

Finally, again based on the definition of the Mordukhovich normal cone, we introduce a concept for the derivative of a general set-valued mapping:

Definition 2.4 Let $\Phi : \mathbb{R}^n \rightrightarrows \mathbb{R}^m$ have a closed graph. Fix any $(\bar{x}, \bar{y}) \in \text{gr}\,\Phi$. Then, the **coderivative** of Φ at $(\bar{x}, \bar{y})$ is defined as a multifunction $D^*\Phi(\bar{x}, \bar{y}) : \mathbb{R}^m \rightrightarrows \mathbb{R}^n$ such that

$$D^*\Phi(\bar{x}, \bar{y})(y^*) := \{x^* | (x^*, -y^*) \in N_{\text{gr}\,\Phi}(\bar{x}, \bar{y})\}$$

One should observe that the coderivative is not just defined at an argument $\bar{x}$ of the preimage space but also needs the specification of a point $\bar{y} \in \Phi(\bar{x})$ in the image $\Phi(\bar{x})$. Indeed, the coderivative is generally a different mapping for different $\bar{y} \in \Phi(\bar{x})$ even for foxed $\bar{x}$. In case of single-valued Φ, one necessarily has $\bar{y} = \Phi(\bar{x})$, so the specification of $\bar{y}$ is omitted and one simply writes $D^*\Phi(\bar{x})$ rather than $D^*\Phi(\bar{x}, f(\bar{x}))$. It can be shown that for single-valued, continuously differentiable mappings Φ the coderivative of Φ at $(\bar{x})$ reduces to its adjoint Jacobian $D^T\Phi(\bar{x})$.

Exercise 2.4 Let $(\bar{x}, \bar{y}) \in \text{gr}\,N_{\mathbb{R}_+}$. Using Example 2.1 for computing $N_{\text{gr}\,N_{\mathbb{R}_+}}(\bar{x}, \bar{y})$, show that

$$\text{If } (\bar{x}, \bar{y}) = (0, 0) \Longrightarrow D^*N_{\mathbb{R}_+}(\bar{x}, \bar{y})(y^*) = \begin{cases} \mathbb{R} & \text{if } y^* = 0 \\ \{0\} & \text{if } y^* > 0 \\ \mathbb{R}_- & \text{if } y^* < 0 \end{cases}$$

$$\text{If } \bar{x} > 0, \ \bar{y} = 0 \Longrightarrow D^*N_{\mathbb{R}_+}(\bar{x}, \bar{y})(y^*) = \{0\}$$

$$\text{If } \bar{x} = 0, \ \bar{y} < 0 \Longrightarrow D^*N_{\mathbb{R}_+}(\bar{x}, \bar{y})(y^*) = \begin{cases} \mathbb{R} & \text{if } y^* = 0 \\ \emptyset & \text{if } y^* \neq 0 \end{cases}$$

The following important scalarization formula for coderivatives of single-valued mappings and subdifferentials of their components holds true:

Proposition 2.1 (see [6], Theorem 1.90) If $\Phi : \mathbb{R}^n \to \mathbb{R}^m$ is locally Lipschitzian, then $D^*\Phi(\bar{x})(y^*) = \partial\langle y^*, \Phi\rangle(\bar{x})$ for all $y^* \in \mathbb{R}^m$.

2.2.2 Lipschitz Properties of Set-Valued Mappings

In this section, we consider a set-valued mapping $F : X \rightrightarrows Y$ between metric spaces, i.e. a mapping assigning to each $x \in X$ an image set $F(x) \subseteq Y$. We want to introduce two generalizations of Lipschitz properties from single-valued to set-valued mappings. We recall that a single-valued mapping $f : X \to Y$ is locally Lipschitzian at some $\bar{x} \in X$ if there exists some $L \geq 0$ such that $d(f(x_1), f(x_2)) \leq Ld(x_1, x_2)$ for all x_1, x_2 in a neighbourhood of $\bar{x}$. A strictly weaker, yet related to Lipschitzian behaviour property results from fixing one of the two arguments in the previous definition. More precisely, f is *calm* at $\bar{x}$, if $d(f(x), f(\bar{x})) \leq L d(x, \bar{x})$ for all x in a neighbourhood of $\bar{x}$. A function which is calm but fails to be locally Lipschitz (due to unbounded slopes for pairs of points close to the fixed point) is illustrated in Fig. 2.4.

When transferring these concepts to set-valued mappings, one has to take into account first that images $F(x)$ are sets now. A straightforward generalization would be obtained by considering the Hausdorff distance between subsets of the image space (see Fig. 2.4):

$$d_H(A, B) := \max\{\sup_{a \in A} d(a, B), \sup_{b \in B} d(b, A)\} \quad \forall A, B \subseteq Y.$$

Then, for instance, local Lipschitz continuity of F at some $\bar{x} \in X$ would amount to the relation $d_H(F(x_1), F(x_2)) \leq L d(x_1, x_2)$ for all x_1, x_2 in a neighbourhood of $\bar{x}$. However, the use of the Hausdorff distance in variational analysis has several drawbacks: first, if the considered sets are unbounded, then convergence of sets may not be well reflected (see Fig. 2.4, where 'intuitively' sets A_n converge to A while $d_H(A_n, A) = \infty$); second, the Hausdorff distance is a global measure and may even for bounded sets not well describe the local convergence of sets around a fixed point in the limit set (see Fig. 2.4).

In order to circumvent the mentioned inconveniences of the use of the Hausdorff distance, the following definitions have proven to be useful for describing the (locally)

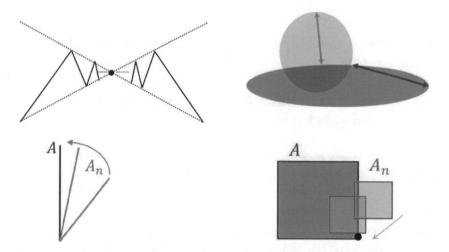

Fig. 2.4 Top left: example for a calm function not being locally Lipschitzian; top right: illustration of the Hausdorff distance as maximum of the excesses of one set over the other; bottom left: 'convergence' of sets having Hausdorff distance ∞ to the limit set; bottom right: the sequence of sets does not converge to the set A in a global sense, although it does so locally around the fixed point

Lipschitzian behaviour of a set-valued mapping $F : X \rightrightarrows Y$ between metric spaces (see, e.g. [8]):

Definition 2.5 F has the **Aubin property** at $(\bar{x}, \bar{y}) \in \mathrm{gr}\, F$ if there are $L, \delta > 0$ with

$$d\,(y, F\,(x_1)) \leq Ld\,(x_1, x_2) \quad \forall x_1, x_2 \in \mathbb{B}_\delta(\bar{x}), \ \forall y \in \left[F\,(x_2) \cap \mathbb{B}_\delta(\bar{y})\right].$$

F is said to be **calm** at $(\bar{x}, \bar{y}) \in \mathrm{gr}\, F$ if there are $L, \delta > 0$ such that

$$d\,(y, F\,(\bar{x})) \leq Ld\,(x, \bar{x}) \quad \forall x \in \mathbb{B}_\delta(\bar{x}), \ \forall y \in \left[F\,(x) \cap \mathbb{B}_\delta(\bar{y})\right].$$

Here, $\mathbb{B}_\delta(z)$ refers to the closed ball around z with radius δ.

One immediately verifies that the Aubin property and calmness presented in Definition 2.5 reduce in the case of single-valued functions to local Lipschitz continuity and calmness as introduced in the beginning of this section. Therefore, it is clear that for set-valued mappings too, calmness is strictly weaker than the Aubin property. Observe also that, in contrast to single-valued mappings, we now have not only to fix some argument $\bar{x} \in X$ but also a point $\bar{y} \in F(\bar{x})$ in the image set, because the local behaviour of F at $(\bar{x}, y)$ may be different for different $y \in F(\bar{x})$.

Exercise 2.5 Show that the mapping $F(t) := \{x \mid x^2 \geq t\}$ is calm but fails to have the Aubin property at the point $(0, 0)$ of its graph.

2.3 Calmness and Aubin Property in Optimization Problems

2.3.1 Calmness as a Constraint Qualification for Abstract Optimization Problems

In this section, we want to derive dual necessary optimality conditions for the abstract optimization problem (2.1). Observe first that (2.1) can be compactly rewritten as $\min\{f(x) \mid x \in G^{-1}(C)\}$. Therefore, we may apply Theorem 2.2 to derive the following

Corollary 2.1 *Let $\bar{x}$ be a local solution of problem (2.1), where we assume that f is locally Lipschitzian, G is continuous, and C is closed. Then, $0 \in \partial f(\bar{x}) + N_{G^{-1}(C)}(\bar{x})$.*

The necessary optimality condition obtained in the Corollary is also referred to as an *abstract M-stationarity condition* because it is based on the Mordukhovich normal cone and the associated subdifferential. As the name suggests, the condition is abstract in that it does not provide yet an expression for the normal cone in terms of the data G and C of problem (2.1). Any condition providing such resolution of normal cones in terms of constraint data is usually called a *constraint qualification*. A key constraint qualification in the context of M-stationarity is calmness as introduced in Definition 2.5. This is explained by the following *preimage formula*:

Theorem 2.3 (see [2], Theorem 4.1) *Let* $G : \mathbb{R}^n \to \mathbb{R}^m$ *be locally Lipschitz, and let* $C \subseteq \mathbb{R}^m$ *be closed. If, for some* $\bar{x} \in G^{-1}(C)$ *the multifunction* $\Psi(y) := \{x | G(x) + y \in C\}$ *is calm at* $(0, \bar{x})$, *then*

$$N_{G^{-1}(C)}(\bar{x}) \subseteq D^* G(\bar{x})\left[N_C(G(\bar{x}))\right].$$

Combining this preimage formula with Corollary 2.1 leads immediately to the following *resolved M-stationarity conditions* for problem (2.1):

Corollary 2.2 *Let* $\bar{x}$ *be a local solution of problem* (2.1), *where we assume that* f *and* G *are locally Lipschitzian and* C *is closed. Under the calmness assumption for* Ψ *in Theorem 2.3, there exists some* $v^* \in N_C(G(\bar{x}))$ *such that* $0 \in \partial f(\bar{x}) + D^* G(\bar{x})(v^*)$.

In the special case that f and G are continuously differentiable and that $C = \mathbb{R}^p_+$, the last corollary yields the classical Karush–Kuhn–Tucker conditions of classical nonlinear optimization under inequality constraints. Indeed, recalling that in this smooth case $\partial f(\bar{x}) = \nabla f(\bar{x})$ and the coderivative coincides with the adjoint Jacobian $D^T G(\bar{x})$, the inclusion from Corollary 2.2 reduces to the equation $0 = \partial f(\bar{x}) + D^T G(\bar{x})v^*$. On the other hand, $v^* \in N_{\mathbb{R}^p_+} G(\bar{x}) = \widehat{N}_{\mathbb{R}^p_+} G(\bar{x})$ by convexity of $\mathbb{R}^p_+$. Now, (2.3) entails the complementarity relations

$$G(\bar{x}) \geq 0, \quad v^* \leq 0, \quad \sum_{i=1}^{p} v_i^* G_i(\bar{x}) = 0.$$

These statements suggest to compare the constraint qualification (CQ) via calmness in Theorem 2.3 with known CQs in nonlinear programming. One can show that calmness implies the *Abadie CQ* but is implied by the *Mangasarian–Fromovitz CQ* (MFCQ) which in turn is equivalent to the stronger Aubin property of Ψ in Theorem 2.3 (see Exercise 2.6). This observation from nonlinear programming already provides some idea of how calmness could work as a strictly weaker CQ in MPECs than the (easier to characterize algebraically) Aubin property.

2.3.2   Verification of Calmness and Aubin Property

As far as the Aubin property of set-valued mapping is concerned, the coderivative introduced in Definition 2.4 provides a powerful equivalent characterization via the celebrated *Mordukhovich criterion*:

Theorem 2.4 (see [8], Theorem 9.40) *Let* $F : \mathbb{R}^n \rightrightarrows \mathbb{R}^m$ *have a closed graph. Then,* F *has the Aubin property at* $(\bar{x}, \bar{y}) \in \mathrm{gr}F$ *if and only if*

$$D^* F(\bar{x}, \bar{y})(0) = \{0\}$$

This criterion may be used, in order to derive an easy-to-verify algebraic characterization of the Aubin property for smooth constraint systems:

Proposition 2.2 *Consider the set-valued mapping given by the perturbation*

$$F(p) := \{x \in \mathbb{R}^n | G(x) - p \in C\}$$

of the constraint $G(x) \in C$ in the abstract optimization problem (2.1). Here, we assume that G is continuously differentiable and C is closed. Let $\bar{x}$ be a feasible point of the unperturbed problem, i.e. $G(\bar{x}) \in C$. Then,

$$F \text{ has the Aubin property at } (0, \bar{x}) \Longleftrightarrow \text{Ker}[\nabla G(\bar{x})]^T \cap N_C(G(\bar{x})) = \{0\}$$

Proof Define $\widetilde{G}(p, x) := G(x) - p$. Then, $\text{gr}F = \widetilde{G}^{-1}(C)$. Clearly, the Jacobian $\nabla \widetilde{G}(0, \bar{x}) = (-I, \nabla G(\bar{x}))$ is surjective. This allows us to invoke the following preimage formula

$$N_{\widetilde{G}^{-1}(C)}(0, \bar{x}) = [\nabla \widetilde{G}(0, \bar{x})]^T N_C(\widetilde{G}(0, \bar{x}))$$

(see [8], [Exercise 6.7]), which in our special case with smooth mappings corresponds to the inclusion of Theorem 2.3 being actually satisfied as an equality. Hence,

$$N_{\text{gr}F}(0, \bar{x}) = N_{\widetilde{G}^{-1}(C)}(0, \bar{x}) = \begin{pmatrix} -I \\ [\nabla G(\bar{x})]^T \end{pmatrix} N_C(\widetilde{G}(0, \bar{x})).$$

Observing that $\widetilde{G}(0, \bar{x}) = G(\bar{x})$, we derive from this last relation that for all p^* the following holds true:

$$(p^*, 0) \in N_{\text{gr}F}(0, \bar{x}) \Longleftrightarrow \exists z^* \in N_C(G(\bar{x})) : \ p^* = -z^*, \ [\nabla G(\bar{x})]^T z^* = 0$$
$$\Longleftrightarrow -p^* \in \text{Ker}[\nabla G(\bar{x})]^T \cap N_C(G(\bar{x})).$$

By Definition 2.4, this amounts to

$$D^* F(0, \bar{x})(0) = \{p^* | (p^*, 0) \in N_{\text{gr}F}(0, \bar{x})\} = - \left\{ \text{Ker}[\nabla G(\bar{x})]^T \cap N_\Theta(G(\bar{x})) \right\}.$$

The result now follows from Theorem 2.4.

Exercise 2.6 For a smooth inequality system $G(x) \in \mathbb{R}^p_-$ the Aubin property of the perturbation mapping $F(p) := \{x \in \mathbb{R}^n | G(x) - p \in \mathbb{R}^p_+\}$ at some feasible point $\bar{x}$ is equivalent to the validity of the Mangasarian–Fromovitz CQ at $\bar{x}$, i.e. with the existence of some d such that

$$\langle \nabla G_i(\bar{x}), d \rangle < 0 \ \ \forall i : G_i(\bar{x}) = 0 \tag{2.6}$$

Hint: using Proposition 2.2 and (2.3) show that F has the Aubin property at $(0, \bar{x})$ if and only if the following relation holds true:

$$[\nabla G(\bar{x})]^T \lambda = 0, \; \lambda \geq 0, \; \lambda_i = 0 \; \forall i : \; G_i(\bar{x}) < 0 \Rightarrow \lambda = 0,$$

which by Motzkin's Theorem of the alternative is equivalent to (2.6).

If it comes to check the calmness property of a set-valued mapping, then one could of course keep using the criterion of Theorem 2.4 for the stronger Aubin property. But in this way, one might loose the potential of strictly weakening the assumptions needed, for instance, for the derivation of necessary optimality conditions. Refining the criterion of Theorem 2.4 towards calmness seems to be possible only in special cases (see [1, Theorem 3.1], [2, Theorem 3.2]). An instance, where calmness (but not necessarily the Aubin property) may always be taken for granted without further assumptions, is *polyhedral mappings*, i.e. set-valued mappings whose graph is a finite union of convex polyhedra. The following result is a slight reduction of a theorem by Robinson:

Proposition 2.3 ([7], Proposition 1) *A polyhedral set-valued mapping is calm at any point of its graph.*

Note that the graph of a polyhedral mapping need not be convex. A prototype example is the set $\operatorname{gr} N_{\mathbb{R}^p_+}$:

Example 2.2 From (2.4), we know that $\operatorname{gr} N_{\mathbb{R}^p_+}$ is the union of two polyhedra:

$$\operatorname{gr} N_{\mathbb{R}_+} = \underbrace{\left[\mathbb{R}_+ \times \{0\} \right]}_{A_0} \cup \underbrace{\left[\{0\} \times \mathbb{R}_- \right]}_{A_1}$$

so that the set Λ from Exercise 2.2 may be represented as a finite union of polyhedra because the Cartesian product of polyhedra is a polyhedron again:

$$\Lambda = \bigcup_{(i_1,\dots,i_p)\in\{0,1\}^p} \underbrace{A_{i_1} \times \cdots \times A_{i_p}}_{\text{polyhedron}}.$$

Now, owing to Exercise 2.2 and the mapping L defined there, we arrive at

$$\operatorname{gr} N_{\mathbb{R}^p_+} = \bigcup_{(i_1,\dots,i_p)\in\{0,1\}^p} \underbrace{L^{-1}(A_{i_1} \times \cdots \times A_{i_p})}_{\text{polyhedron}}.$$

showing that $\operatorname{gr} N_{\mathbb{R}^p_+}$ is a polyhedron as the preimage of a polyhedron under a linear mapping. Consequently, the normal cone mapping $x \mapsto N_{\mathbb{R}^p_+}$ is polyhedral.

In many applications, set-valued mappings are neither polyhedral nor satisfy the Aubin property, so that the previous approaches for verifying calmness would not apply. On the other hand, more than often a structure is present which is partially polyhedral and partially 'Aubin-like'. In such cases, the following useful characterization of structured calmness can be useful:

Theorem 2.5 ([5], Theorem 3.6) *Let* $T_1 : X_1 \rightrightarrows X$ *and* $T_2 : X_2 \rightrightarrows X$ *be multifunctions between metric spaces* X_1, X_2, X. *If*

1. T_1 *is calm at* $(x_1, x) \in \operatorname{gr} T_1$
2. T_2 *is calm at* $(x_2, x) \in \operatorname{gr} T_2$
3. T_2^{-1} *has the Aubin property at* (x, x_2)
4. $T_1(x_1) \cap T_2(\cdot)$ *is calm at* (x_2, x),

then the multifunction $(T_1 \cap T_2)(x_1, x_2) := T_1(x_1) \cap T_2(x_2)$ *is calm at* (x_1, x_2, x).

Exercise 2.7 Provide an example for two set-valued mappings being calm but without their pointwise intersection being calm likewise.

The previous exercise illustrates, why the first two conditions of Theorem 2.5 alone are not sufficient to yield the calmness of the intersection mapping.

2.4 M-Stationarity Conditions for MPECs

We consider the MPEC introduced in (2.2) and specify now that the normal cone N appearing there refers to the Mordukhovich normal cone, so for convex sets Γ it coincides with the normal cone of convex analysis. On the other hand, the use of the Mordukhovich normal cone allows an application to general closed sets and, in contrast to the Fréchet normal cone, its graph will be always closed (see Sect. 2.2.1). Observe that by passing to the concept of the graph of a multifunction, one may equivalently rewrite it as

$$\min\{\varphi(x, y) \mid \underbrace{(y, -F(x, y))}_{H(x,y)} \in \operatorname{gr} N_\Gamma\}, \tag{2.7}$$

which is exactly of the form of (2.1) with $f := \varphi$, $G := H$ [(as defined in (2.7)] and $C := \operatorname{gr} N_\Gamma$. This being done, we may immediately apply Corollary 2.2 on M-stationarity conditions for abstract optimization problems in order to specify them in the case of MPECs:

Proposition 2.4 *Let* $(\bar{x}, \bar{y})$ *be a local solution of the MPEC* (2.2), *where we assume that* φ *and* F *are locally Lipschitzian and* Γ *is closed. Then, if the mapping*

$$\Psi(p_1, p_2) := \{(x, y) \mid p_2 \in F(x, y) + N_\Gamma(y + p_1)\} \tag{2.8}$$

is calm at $(0, 0, \bar{x}, \bar{y})$, *then there exists an MPEC multiplier* $v^* \in N_{\operatorname{gr} N_\Gamma}(H(\bar{x}, \bar{y}))$ *(with H as introduced in (2.7)) such that*

$$0 \in \partial \varphi(\bar{x}, \bar{y}) + D^*(H(\bar{x}, \bar{y}))(v^*). \tag{2.9}$$

Proof We consider the MPEC (2.2) in its equivalent form (2.7). Since F is locally Lipschitzian, the same holds true for H. Moreover, as mentioned above, the set $C :=$ gr N_Γ is closed. Finally, we observe that the calmness assumption in our proposition amounts in graphical form to the calmness of the mapping

$$\Psi(p_1, p_2) := \{(x, y) | (p_1 + y, p_2 - F(x, y)) \in \text{gr } N_\Gamma\}$$

at $(0, 0, \bar{x}, \bar{y})$. Recalling the definition of H and putting $p := (p_1, p_2)$ this can be rephrased as the calmness of the mapping $\Psi(p) := \{(x, y) | (p + H(x, y)) \in \text{gr } N_\Gamma\}$ occurring in Theorem 2.3 and needed in Corollary 2.2. Summarizing, all assumptions of Corollary 2.2 are satisfied for problem (2.7) and the assertion follows.

The necessary optimality condition (2.9) is not fully efficient, yet in that it is formulated in terms of the intermediary mapping H rather than the input mapping F of (2.7). Moreover, one can simplify the calmness condition according to the following statement.

Lemma 2.1 (see [9], Proposition 5.2) *The full perturbation mapping Ψ in (2.8) is calm at $(0, 0, \bar{x}, \bar{y})$ if and only if the associated reduced perturbation mapping $\tilde{\Psi}(p) := \{(x, y) | p \in F(x, y) + N_\Gamma(y)\}$ is calm at $(0, \bar{x}, \bar{y})$.*

Next, we develop a more handy expression for the coderivative of $H(x, y) = (y, -F(x, y))$ in (2.9). Put $H := (H_1, H_2)$. The scalarization formula of Proposition 2.1 and the sum rule of Theorem 2.1 yield that

$$
\begin{aligned}
D^*H(\bar{x}, \bar{y})(u^*, v^*) &= \partial \langle (u^*, v^*), (H_1, H_2) \rangle (\bar{x}, \bar{y}) \\
&\subseteq \partial \langle u^*, H_1 \rangle (\bar{x}, \bar{y}) + \partial \langle v^*, H_2 \rangle (\bar{x}, \bar{y}) \\
&= \{(0, u^*)\} + D^*H_2(\bar{x}, \bar{y})(v^*) \\
&= \{(0, u^*)\} + D^*(-F)(\bar{x}, \bar{y})(v^*). \tag{2.10}
\end{aligned}
$$

Here, we made use of the fact that $\langle u^*, H_1 \rangle = \langle u^*, y \rangle$ is a linear function of (x, y) and, hence, the subdifferential reduces to its gradient. Combining (2.10) with Lemma 2.1, Proposition 2.4 yields the following M-stationarity conditions for the MPEC (2.2) with locally Lipschitzian mappings completely in terms of the input data of the problem:

Theorem 2.6 *Let $(\bar{x}, \bar{y})$ be a local solution of the MPEC (2.2), where φ and F are locally Lipschitzian and Γ is closed. Then, if the mapping $\tilde{\Psi}$ in Lemma 2.1 is calm at $(0, \bar{x}, \bar{y})$, then there exist MPEC multipliers $(u^*, v^*) \in N_{\text{gr}N_\Gamma}(\bar{y}, -F(\bar{x}, \bar{y}))$ such that*

$$0 \in \partial \varphi(\bar{x}, \bar{y}) + \{(0, u^*)\} + D^*(-F)(\bar{x}, \bar{y})(v^*).$$

In the following, we are going to specify the M-stationarity conditions of Theorem 2.6 to the case of smooth input data for the MPEC (2.2):

Corollary 2.3 *Let $(\bar{x}, \bar{y})$ be a local solution of the MPEC (2.2), where φ and F are continuously differentiable and Γ is closed. Then, if the mapping $\widetilde{\Psi}$ in Lemma 2.1 is calm at $(0, \bar{x}, \bar{y})$, then there exists an MPEC multiplier v^* such that*

$$0 = \nabla_x \varphi(\bar{x}, \bar{y}) + \left[\nabla_x F(\bar{x}, \bar{y})\right]^T v^*$$
$$0 \in \nabla_y \varphi(\bar{x}, \bar{y}) + \left[\nabla_y F(\bar{x}, \bar{y})\right]^T v^* + D^* N_\Gamma(\bar{y}, -F(\bar{x}, \bar{y}))(v^*).$$

Proof In the case of smooth data, the subdifferential and coderivative, respectively, reduce to

$$\partial \varphi(\bar{x}, \bar{y}) = (\nabla_x \varphi(\bar{x}, \bar{y}), \nabla_y \varphi(\bar{x}, \bar{y}))$$
$$D^*(-F)(\bar{x}, \bar{y})(v^*) = (-\left[\nabla_x F(\bar{x}, \bar{y})\right]^T v^*, -\left[\nabla_y F(\bar{x}, \bar{y})\right]^T v^*).$$

Now, Theorem 2.6 guarantees the existence of $(u^*, v^*) \in N_{\mathrm{gr} N_\Gamma}(\bar{y}, -F(\bar{x}, \bar{y}))$ such that

$$0 = \nabla_x \varphi(\bar{x}, \bar{y}) - \left[\nabla_x F(\bar{x}, \bar{y})\right]^T v^*$$
$$0 = \nabla_y \varphi(\bar{x}, \bar{y}) - \left[\nabla_y F(\bar{x}, \bar{y})\right]^T v^* + u^*$$

Since $(u^*, v^*) \in N_{\mathrm{gr} N_\Gamma}(\bar{y}, -F(\bar{x}, \bar{y}))$ if and only if $u^* \in D^* N_\Gamma(\bar{y}, -F(\bar{x}, \bar{y}))(-v^*)$ (see Definition 2.4), we can substitute for the second multiplier u^* by turning the second equation above into an inclusion.

We note that the preceding Corollary has been proven first in [10, Theorem 3.2] using a different way of reasoning. Looking at Corollary 2.3, there remain two issues to be clarified for an efficient application of the obtained M-stationarity conditions: first, the calmness of the perturbation mapping $\widetilde{\Psi}$ has to be verified, and second, explicit formulae for the coderivative $D^* N_\Gamma$ have to be found. This will be the object of the following two sections.

2.5 Verification of Calmness for the Perturbation Mapping

2.5.1 Using the Aubin Property

We start by providing an algebraic condition ensuring the calmness of the mapping $\widetilde{\Psi}$ introduced in Lemma 2.1 via checking the stronger Aubin property for the perturbation mapping (2.8). Observe first the general relation $\widetilde{\Psi}(p_2) = \Psi(0, p_2)$ between both mappings. This means that the perturbation of Ψ is richer than that of $\widetilde{\Psi}$ while $\widetilde{\Psi}(0) = \Psi(0, 0)$. As a consequence, Ψ having the Aubin property at $(0, 0, \bar{x}, \bar{y})$ would imply $\widetilde{\Psi}$ having the Aubin property at $(0, \bar{x}, \bar{y})$ which in turn would imply

$\widetilde{\Psi}$ being calm at $(0, \bar{x}, \bar{y})$. Therefore, the following proposition yields a sufficient algebraic condition for the calmness of $\widetilde{\Psi}$ as required in Corollary 2.3:

Proposition 2.5 *Let F be continuously differentiable, let Γ be closed, and let $(\bar{x}, \bar{y})$ be such that $0 \in F(\bar{x}, \bar{y}) + N_\Gamma(\bar{y})\}$. Then, the perturbation mapping Ψ defined in (2.8) has the Aubin property at $(0, 0, \bar{x}, \bar{y})$ if and only if the following implication holds true:*

$$\left[\nabla_x F(\bar{x}, \bar{y})\right]^T v^* = 0, \quad \left[\nabla_y F(\bar{x}, \bar{y})\right]^T v^* \in D^* N_\Gamma(\bar{y}, -F(\bar{x}, \bar{y}))(-v^*) \Longrightarrow v^* = 0$$

As a consequence, this implication guarantees the calmness of $\widetilde{\Psi}$ at $(0, \bar{x}, \bar{y})$ as required in Corollary 2.3.

Proof By (2.8), Ψ may be rewritten in graphical form as

$$\Psi(p_1, p_1) := \{(x, y) | H(x, y) - (p_1, p_1) \in \operatorname{gr} N_\Gamma\},$$

where H is defined in (2.7). Now, by Proposition 2.2, Ψ has the Aubin property at $(0, 0, \bar{x}, \bar{y})$ if and only if

$$\operatorname{Ker} \begin{pmatrix} 0 - \left[\nabla_x F(\bar{x}, \bar{y})\right]^T \\ I - \left[\nabla_y F(\bar{x}, \bar{y})\right]^T \end{pmatrix} \cap N_{\operatorname{gr} N_\Gamma}(\bar{y}, -F(\bar{x}, \bar{y})) = \{0\}.$$

which is equivalent to the implication

$$\left[\nabla_x F(\bar{x}, \bar{y})\right]^T v^* = 0, \quad u^* - \left[\nabla_y F(\bar{x}, \bar{y})\right]^T v^* = 0, \quad u^* \in D^* N_\Gamma(\bar{y}, -F(\bar{x}, \bar{y}))(-v^*)$$
$$\Longrightarrow u^* = 0, \ v^* = 0.$$

This yields the assertion.

Of course, the application of Proposition 2.5 hinges on concrete formulae for the coderivative $D^* N_\Gamma$. Possibilities to do so will be discussed in Sect. 2.6. Alternatively, one could try to check the Aubin property of $\widetilde{\Psi}$ directly using the definition in order to deduce its calmness. This is illustrated in the following example:

Example 2.3 Let $F : \mathbb{R} \times \mathbb{R}^2 \to \mathbb{R}^2$ be given by $F(x, y_1, y_2) := (0, 1)$ and

$$\Gamma := \{y \in \mathbb{R}^2 | y_2 \geq 0, \ y_2 \geq y_1^2\}.$$

Then, $\widetilde{\Psi}$ introduced in Lemma 2.1 takes the form

$$\widetilde{\Psi}(p_1, p_2) = \{(x, y_1, y_2) | (p_1, p_2 - 1) \in N_\Gamma(y)\}$$

We verify the Aubin property of $\widetilde{\Psi}$ at the point $(\bar{p}_1, \bar{p}_2, \bar{x}, \bar{y}_1, \bar{y}_2) := (0, 0, 0, 0, 0)$ which belongs to gr $\widetilde{\Psi}$ due to $(0, -1) \in N_\Gamma(y)$. If $y \in \operatorname{int} \Gamma$, then $N_\Gamma(y_1, y_2) =$

$\{(0, 0)\}$. Therefore, $(y_1, y_2) \in \text{bd}\ \Gamma$ for all $(x, y_1, y_2) \in \widetilde{\Psi}(p_1, p_2)$ and (p_1, p_2) close to $(\bar{p}_1, \bar{p}_2)$. In particular, $y_2 = y_1^2$. The first inequality in the definition of Γ has been arranged to be redundant. Hence, $N_\Gamma(y) = \mathbb{R}_+\{(2y_1, -1)\}$ for all $(y_1, y_2) \in \text{bd}\ \Gamma$. This implies that

$$(p_1, p_2 - 1) = \lambda(y)(2y_1, -1)$$

with some function $\lambda(y) \geq 0$ for all (p_1, p_2) close to $(\bar{p}_1, \bar{p}_2)$ and $(x, y_1, y_2) \in \widetilde{\Psi}(p_1, p_2)$. A comparison of components along with $y_2 = y_1^2$ yields the relations

$$\lambda(y) = 1 - p_2, \quad y_1 = p_1/2(1 - p_2), \quad y_2 = (p_1/2(1 - p_2))^2.$$

Consequently, for p close to $\bar{p}$ we have that

$$\widetilde{\Psi}(p) = \{(x, y)|y_1 = p_1/2(1 - p_2), \ y_2 = (p_1/2(1 - p_2))^2\}.$$

Clearly, the images of $\widetilde{\Psi}$ do not involve x. Moreover, the y-components are locally Lipschitzian functions of (p_1, p_2) in a neighbourhood of $(\bar{p}_1, \bar{p}_2)$. Therefore, $\widetilde{\Psi}$ has the Aubin property at $(\bar{p}_1, \bar{p}_2, \bar{x}, \bar{y}_1, \bar{y}_2)$.

2.5.2   Using Polyhedrality or Structured Calmness

If the MPEC (2.2) is governed by a linear generalized equation (i.e. F is affine linear and Γ is a convex polyhedron), then calmness of the perturbation mapping $\widetilde{\Psi}$ introduced in Lemma 2.1 comes for free and, hence, the M-stationarity conditions of Corollary 2.3 can be derived without any further assumptions. More precisely, we have the following result:

Proposition 2.6 *If in the MPEC* (2.2) *(or* (2.7)*, respectively)* Γ *is a polyhedron and* $F(x, y) = Ax + By + c$ *is an affine linear mapping, then the perturbation mapping* $\widetilde{\Psi}$ *introduced in Lemma 2.1 is calm at all points of its graph.*

Proof By definition of $\widetilde{\Psi}$, it holds that

$$\text{gr}\ \widetilde{\Psi} = \{(p, x, y)|\ \underbrace{(y, p - Ax - By - c)}_{H(p,x,y)} \in \text{gr}\ N_\Gamma\} = H^{-1}(\text{gr}\ N_\Gamma)$$

Since H is an affine linear mapping, it will be sufficient to verify that $\text{gr}\ N_\Gamma$ is a finite union of polyhedra, because then so is $\text{gr}\ \widetilde{\Psi}$ and the result follows from Proposition 2.3.

In order to carry out this verification, we describe the polyhedron Γ explicitly as the solution of a finite inequality system: $\Gamma = \{y | Cy \leq d\}$. It is well known that the normal cone to a polyhedron of such description calculates as

$$N_\Gamma = C^T N_{\mathbb{R}^p_-}(Cy - d).$$

It follows that

$$(y, z) \in \operatorname{gr} N_\Gamma \Leftrightarrow z \in N_\Gamma(y) \Leftrightarrow \exists \lambda \in N_{\mathbb{R}^p_-}(Cy - d) : z = C^T \lambda.$$

Therefore, $\operatorname{gr} N_\Gamma = P(\Theta)$ with $P(y, z, \lambda) := (y, z)$ and

$$\Theta := \{(y, z, \lambda) | z = C^T \lambda, \ (\lambda, Cy - d) \in \operatorname{gr} N_{\mathbb{R}^p_+}\}.$$

Here, we exploited that

$$x^* \in N_{\mathbb{R}^p_-}(x) \Leftrightarrow x \in N_{\mathbb{R}^p_+}(x^*). \tag{2.11}$$

Defining

$$H(y, z, \lambda) := (z - C^T \lambda, \lambda, Cy - d),$$

we then have that $\Theta = H^{-1}\left(\{0\} \times \operatorname{gr} N_{\mathbb{R}^p_+}\right)$. Thus, by Example 2.2, Θ is the preimage of a finite union of polyhedra under a linear mapping and as such is a finite union $\Theta = \cup_{i=1}^q A_i$ of certain polyhedra A_i. But then,

$$\operatorname{gr} N_\Gamma = P(\cup_{i=1}^q A_i) = \cup_{i=1}^q P(A_i).$$

As the projection of a polyhedron is a polyhedron again, Γ is a polyhedral map.

Finally, we give an idea about how to employ structured calmness by formulating without proof a result which can be derived from Theorem 2.5 along the lines of [3, Theorem 7.1]:

Theorem 2.7 *For the mapping $\widetilde{\Psi}(p)$ defined in Lemma 2.1 fix any $(\bar{x}, \bar{y}) \in \widetilde{\Psi}(0)$. Assume that Γ is polyhedral and*

$$F(x, y) = \begin{pmatrix} F_1(x, y) \\ F_2(y) \end{pmatrix} \quad \text{with } F_2 \text{ affine linear and } \nabla_x F_1(\bar{x}, \bar{y}) \text{ surjective.}$$

Then, $\widetilde{\Psi}$ is calm at $(0, \bar{x}, \bar{y})$.

Note that in this theorem, neither the stronger Aubin property (or its equivalent characterization via Proposition 2.5) nor the (full) affine linearity of the mapping F as in Proposition 2.6 is required.

2.6   Coderivative Formulae and Fully Explicit
M-Stationarity Conditions

After providing various possibilities of verifying the calmness property of the perturbation mapping, the missing link for an efficient application of Corollary 2.3 is
the computation of the coderivative to the normal cone mapping of the set Γ. Several
results in this direction are known. We content ourselves here with the case of Γ
being described by a finite system of smooth inequalities, i.e.

$$\Gamma := \{y \in \mathbb{R}^m | g_i(y) \le 0 \ (i = 1, \ldots, p)\} \tag{2.12}$$

where the $g = (g_i)$ is a twice continuously differentiable. As unbinding constraints
are of no interest for a local analysis, we assume without loss of generality that
$g(\bar{y}) = 0$ at some fixed point of interest $\bar{y} \in \Gamma$. The following result is a consequence
of a chain rule for second-order subdifferentials presented in [6, Theorem 1.127]:

Theorem 2.8 *In* (2.12), *fix any* $\bar{y} \in \Gamma$ *and* $\bar{v} \in N_\Gamma(\bar{y})$. *Assume that* $g(\bar{y}) = 0$. *If*
$\nabla g(\bar{y})$ *is surjective, then*

$$D^* N_\Gamma(\bar{y}, \bar{v})(v^*) = \left(\sum_{i=1}^p \bar{\lambda}_i \nabla^2 g_i(\bar{y}) \right) v^* + \left[\nabla g(\bar{y}) \right]^T D^* N_{\mathbb{R}^p_-} \left(g(\bar{y}), \bar{\lambda} \right) \left(\nabla g(\bar{y}) v^* \right).$$

Here, $\bar{\lambda}$ *is the unique multiplier satisfying* $\bar{v} = \nabla^T g(\bar{y})\lambda$.

Given Theorem 2.8, the last step needed for calculating the coderivative of N_Γ consists in specifying the coderivative of $N_{\mathbb{R}^p_-}$. We will only need the formula evaluated
at points $(0, \bar{y})$ here.

Lemma 2.2 *Let* $\bar{y} \in N_{\mathbb{R}^p_-}(0)$. *Then,*

$$D^* N_{\mathbb{R}^p_-}(0, \bar{y})(y^*) = \emptyset \quad \text{if there exists some } i \text{ such that } \bar{y}_i > 0, \ y_i^* \ne 0$$

Otherwise:

$$D^* N_{\mathbb{R}^p_-}(0, \bar{y})(y^*) = \left\{ x^* \left| \begin{array}{l} x_i^* = 0 \text{ if } \bar{y}_i = 0, \ y_i^* < 0 \\ x_i^* \ge 0 \text{ if } \bar{y}_i = 0, \ y_i^* > 0 \end{array} \right. \right\}.$$

Proof From (2.11) and Exercise 2.2, we conclude that $\operatorname{gr} N_{\mathbb{R}^p_-} = \widetilde{L}^{-1}(\Lambda)$, where

$$\widetilde{L}(x_1, \ldots, x_p, y_1, \ldots, y_p) := (y_1, x_1, \ldots, y_p, x_p), \quad \Lambda := \operatorname{gr} N_{\mathbb{R}_+} \times \cdots \times \operatorname{gr} N_{\mathbb{R}_+}.$$

Clearly, $\widetilde{L}$ is surjective (actually regular) and $\widetilde{L} = \widetilde{L}^{-1} = \widetilde{L}^T$. Therefore, as in the
proof of Proposition 2.2, we are allowed to invoke the preimage formula from [8,
Exercise 6.7] in order to verify that

$$N_{\operatorname{gr} N_{\mathbb{R}^p_-}}(x, y) = N_{\widetilde{L}^{-1}(\Lambda)}(x, y) = \widetilde{L} N_\Lambda(\widetilde{L}(x, y)).$$

Hence, by (2.5),

$$(x^*, y^*) \in N_{\text{gr } N_{\mathbb{R}^p_-}}(x, y)$$

$$\iff \widetilde{L}^{-1}(x^*, y^*) \in N_\Lambda(\widetilde{L}(x, y))$$

$$\iff (y_1^*, x_1^*, \ldots, y_p^*, x_p^*) \in N_{\text{gr } N_{\mathbb{R}_+}}(y_1, x_1) \times \cdots \times N_{\text{gr } N_{\mathbb{R}_+}}(y_p, x_p).$$

In other words, $x^* \in D^* N_{\mathbb{R}^p_-}(0, \bar{y})(y^*)$ if and only if $-y_i^* \in D^* N_{\mathbb{R}_+}(\bar{y}_i, 0)(-x_i^*)$ for all $i = 1, \ldots p$. Now, using Exercise 2.4, one arrives at the asserted formula: Indeed, to see it for example for the first statement, let there exist some i such that $\bar{y}_i > 0$, $y_i^* \neq 0$. If there was some $x^* \in D^* N_{\mathbb{R}^p_-}(0, \bar{y})(y^*)$, then $-y_i^* \in D^* N_{\mathbb{R}_+}(\bar{y}_i, 0)(-x_i^*)$. Then, the second case in Exercise 2.4 yields the contradiction $y_i^* = 0$. Hence, we infer the desired statement $D^* N_{\mathbb{R}^p_-}(0, \bar{y})(y^*) = \emptyset$. The second asserted statement follows similarly.

The finally obtained coderivative formula allows us to combine Theorem 2.8 with Lemma 2.2 in order to make the M-stationarity conditions of Corollary 2.3 fully explicit:

Theorem 2.9 *Let* $(\bar{x}, \bar{y})$ *be a local solution to the MPEC with smooth data*

$$\min\{\varphi(x, y)|0 \in F(x, y) + N_\Gamma(y)\}, \quad \Gamma := \{y \in \mathbb{R}^p | g_i(y) \leq 0 \ (i = 1, \ldots, p)\},$$

where φ, F *are once and* g *is twice continuously differentiable. Assume that* $g(\bar{y}) = 0$, *that* $\nabla g(\bar{y})$ *is surjective and that the perturbation mapping* $\widetilde{\Psi}$ *from Lemma 2.1 is calm at* $(0, \bar{x}, \bar{y})$. *Then, there exist MPEC multipliers* u^*, v^* *such that*

$$0 = \nabla_x \varphi(\bar{x}, \bar{y}) + \left[\nabla_x F(\bar{x}, \bar{y})\right]^T v^*$$

$$0 = \nabla_y \varphi(\bar{x}, \bar{y}) + \left(\left[\nabla_y F(\bar{x}, \bar{y})\right]^T + \sum_{i=1}^p \bar{\lambda}_i \nabla^2 g_i(\bar{y})\right) v^* + \left[\nabla g(\bar{y})\right]^T u^*$$

$$0 = \nabla g_i(\bar{y}) v^* \quad \forall i : \bar{\lambda}_i > 0$$

$$0 = u_i^* \quad \forall i : \bar{\lambda}_i = 0, \ \nabla g_i(\bar{y}) v^* < 0$$

$$0 \leq u_i^* \quad \forall i : \bar{\lambda}_i = 0, \ \nabla g_i(\bar{y}) v^* > 0$$

Here, $\bar{\lambda}$ *is the unique solution of* $F(\bar{x}, \bar{y}) = \left[\nabla g(\bar{y})\right]^T \bar{\lambda}$.

Proof By Corollary 2.3, there exist multipliers w^*, v^* such that the first of our asserted equations and

$$0 = \nabla_y \varphi(\bar{x}, \bar{y}) - \left[\nabla_y F(\bar{x}, \bar{y})\right]^T v^* + w^*, \quad w^* \in D^* N_\Gamma(\bar{y}, -F(\bar{x}, \bar{y}))(v^*)$$

hold true. By Theorem, 2.8, there exists some $u^* \in D^* N_{\mathbb{R}^p_-} \left(g(\bar{y}), \bar{\lambda} \right) \left(\nabla g(\bar{y}) v^* \right)$ with

$$w^* = \left(\sum_{i=1}^{p} \bar{\lambda}_i \nabla^2 g_i(\bar{y}) \right) v^* + \left[\nabla g(\bar{y}) \right]^T u^*$$

yielding the second of the asserted equations. Now, Lemma 2.2 provides the last asserted relations of Theorem upon recalling that $g(\bar{y}) = 0$ and that

$$u^* \in D^* N_{\mathbb{R}^p_-} \left(g(\bar{y}), \bar{\lambda} \right) \left(\nabla g(\bar{y}) v^* \right) \neq \emptyset.$$

In the special case of a polyhedral set Γ, the surjectivity condition $\nabla g(\bar{y})$ in Theorem 2.8 can be dispensed with and a precise coderivative formula is available too. More precisely, let $\Gamma := \{x \in \mathbb{R}^n | Ax \leq b\}$ for some (q, n)-matrix A. Denote the rows of A by a_i. Fix $\bar{x} \in \Gamma$ and $\bar{v} \in N_\Gamma(\bar{x})$, i.e. $\bar{v} = A^T \lambda$ for some $\lambda \in \mathbb{R}^q_+$. For each $x \in \Gamma$ let $I(x) := \{i | a_i x = b_i\}$. Define the family of active index sets as

$$\mathscr{I} := \{I \subseteq \{1, \ldots, q\} | \exists x \in \Gamma : I = I(x)\}$$

Then, the following coderivative formula holds true for N_Γ:

Theorem 2.10 ([4], Proposition 3.2)

$$D^* N_\Gamma(\bar{x}, \bar{v})(v^*) = \left\{ x^* \left| (x^*, -v^*) \in \bigcup_{J \subseteq I_1 \subseteq I_2 \subseteq I(\bar{x})} P_{I_1, I_2} \times Q_{I_1, I_2} \right. \right\},$$

Here,

$$P_{I_1, I_2} = \text{con} \{a_i | i \in \chi(I_2) \setminus I_1\} + \text{span} \{a_i | i \in I_1\}$$
$$Q_{I_1, I_2} = \{h \in \mathbb{R}^n | \langle a_i, h \rangle = 0 \quad (i \in I_1), \quad \langle a_i, h \rangle \leq 0 \quad (i \in \chi(I_2) \setminus I_1)\}$$

and 'con' and 'span' refer to the convex conic and linear hulls, respectively. Moreover,

$$J := \{j \in I | \lambda_j > 0\} \quad and \quad \chi(I') := \cap \{J \in \mathscr{I} | I' \subseteq J\} \quad \forall I' \subseteq \{1, \ldots, q\}.$$

This last theorem can be combined, for instance, with Proposition 2.6 in order to derive fully explicit M-stationarity conditions in the spirit of Theorem 2.9 without any further assumptions (with the perturbation mapping $\widetilde{\Psi}$ being automatically calm thanks to Proposition 2.6).

References

1. Henrion, R., Outrata, J.: A subdifferential criterion for calmness of multifunctions. J. Math. Anal. Appl. **258**, 110–130 (2001)
2. Henrion, R., Jourani, A., Outrata, J.V.: On the calmness of a class of multifunctions. SIAM J. Optim. **13**, 603–618 (2002)
3. Henrion, R., Outrata, J., Surowiec, T.: Analysis of M-stationary points to an EPEC modeling oligopolistic competition in an electricity spot market. ESAIM Control Optim. Calc. Var. **18**, 295–317 (2012)
4. Henrion, R., Römisch, W.: On M-stationary points for a stochastic equilibrium problem under equilibrium constraints in electricity spot market modeling. Appl. Math. **52**, 473–494 (2007)
5. Klatte, D., Kummer, B.: Constrained minima and Lipschitzian penalties in metric mpaces. SIAM J. Optim. **13**, 619–633 (2002)
6. Mordukhovich, B.S.: Variational Analysis and Generalized Differentiation. Springer, Berlin (2006)
7. Robinson, S.M.: Some continuity properties of polyhedral multifunctions. Math. Program. Stud. **14**, 206–214 (1976)
8. Rockafellar, R.T., Wets, J.-B.: Variational Analysis. Springer, Berlin (1998)
9. Surowiec, T.: Explicit stationarity conditions and solution characterization for equilibrium problems with equilibrium constraints. Ph.D. thesis, Humboldt University Berlin (2010)
10. Ye, J.J., Ye, X.Y.: Necessary optimality conditions for optimization problems with variational inequality constraints. Math. Oper. Res. **22**, 977–997 (1997)

Chapter 3
Optimality Conditions for Bilevel Programming: An Approach Through Variational Analysis

Joydeep Dutta

Abstract In this article, we focus on the study of bilevel programming problems where the feasible set in the lower-level problem is an abstract closed convex set and not described by any equalities and inequalities. In such a situation, we can view them as MPEC problems and develop necessary optimality conditions. We also relate various solution concepts in bilevel programming and establish some new connections. We study in considerable detail the notion of partial calmness and its application to derive necessary optimality conditions and also give some illustrative examples.

3.1 Introduction

In this article, we focus on optimality conditions for bilevel programming. We shall consider a bilevel programming problem in the following form

$$\min_{x} F(x, y), \quad \text{subject to} \quad x \in X, y \in S(x),$$

where for each $x \subset X$ the set $S(x)$ is given as

$$S(x) = argmin_y\{f(x, y) : y \in K(x)\},$$

where $F : \mathbb{R}^n \times \mathbb{R}^m \to \mathbb{R}, f : \mathbb{R}^n \times \mathbb{R}^m \to \mathbb{R}$ and $K(x)$ is a closed convex set in $\mathbb{R}^m$ depending on $x \in X$.

J. Dutta (✉)
Economics Group, Department of Humanities and Social Sciences,
Indian Institute of Technology, Kanpur, Kanpur 208016, India
e-mail: jdutta@iitk.ac.in

© Springer Nature Singapore Pte Ltd. 2017   43
D. Aussel and C. S. Lalitha (eds.), *Generalized Nash Equilibrium
Problems, Bilevel Programming and MPEC*, Forum for Interdisciplinary
Mathematics, https://doi.org/10.1007/978-981-10-4774-9_3

For simplicity of the presentation throughout our presentation, we shall restrict ourselves to the case where $X = \mathbb{R}^n$. Thus, the bilevel programming problem now looks like.

$$\min_x F(x, y), \quad \text{subject to} \quad y \in S(x),$$

where for each $x \in \mathbb{R}^n$ the set $S(x)$ is given as

$$S(x) = argmin_y\{f(x, y) : y \in K(x)\}.$$

We shall denote this bilevel programming above as (BP). We shall make the following assumptions

(i) $F(x, y)$ is locally Lipschitz in both variables.
(ii) For each $x \in \mathbb{R}^n$, the function $y \mapsto f(x, y)$ is convex in y.
(iii) $K(x)$ is a closed convex for each $x \in \mathbb{R}^n$.

If for each $x \in \mathbb{R}^n$, the solution set $S(x)$ is singleton given as $S(x) = \{y(x)\}$, then the bilevel programming problem (BP) can be posed as

$$\min_x F(x, y(x))$$

However, this is not the case if $S(x)$ fails to be singleton even just for a some finite number $x \in \mathbb{R}^n$. In such a case where $S(x)$ is not singleton for each x, one defines two different types of bilevel problem, namely optimistic problem and the pessimistic problem. The need to have these two separate versions springs from the fact that if the solution map S is not singleton then the bilevel programming problem can be expressed as the following set-valued optimization problem

$$\min_x F(x, S(x)) = \bigcup_{y \in S(x)} F(x, y).$$

Though there is a vast literature on set-valued optimization (see, for example, Jahn [15]), it seems that the methods of set-valued optimization are not at present adequate to address the concerns of bilevel programming. This leads us to develop two types of bilevel problem in order to have a single-valued objective at the upper level. The first one is the *optimistic formulation* which is given as follows. Consider that $S(x) \neq \emptyset$ for each x, and define the function

$$\varphi_0(x) = \min_{y \in S(x)} F(x, y).$$

Then, the *optimistic problem* is to minimize φ_0 over x. We shall refer to the optimistic problems as (BP_o). Thus, the optimistic problem is viewed as as bilevel problem where the follower cooperates with the leader.

For the pessimistic problem, one defines the function

$$\varphi_p(x) = \max_{y \in S(x)} F(x, y).$$

Thus, the pessimistic bilevel problem consists of minimizing φ_p over $\mathbb{R}^n$. Note that the pessimistic formulation of a bilevel problem is viewed as one where the follower does not cooperate with the leader. The optimistic problem though somewhat manageable the pessimistic problem seems to be a hard nut to crack. However, very recently some progress has been made in this direction in Dempe, Mordukhovich and Zemkoho [10]. In this article, we shall, however, focus on the optimistic bilevel programming problem. A more tractable version of an optimistic bilevel programming is the following problem where the upper-level optimization problem is minimized in terms of both x and y. This problem which is denoted as (OBP) is given as

$$\min_{x,y} F(x, y), \quad \text{subject to} \quad y \in S(x).$$

To the best of our knowledge, the problem (OBP) was first analysed in Ye and Zhu [21]. It was also studied, for example, in Ye [22], Ye and Zhu [24], Dutta and Dempe [11] and termed as the optimistic bilevel programming in Dempe, Dutta and Mordukhovich [8]. The justification for calling (OBP) the optimistic bilevel problem stems from the close relationship it has with the problem (BP_o). Before we proceed, we would like to note that the relation between (OBP) and (BP_o) was analysed, for example, in Dutta and Dempe [11]. We discuss this link afresh under much milder conditions than in [11].

Proposition 3.1 *Let $\bar{x}$ be the local solution of the optimistic formulation (BP_o). Then for any $\bar{y} \in S(\bar{x})$, the vector $(\bar{x}, \bar{y})$ is a local minimum of (OBP) if $\bar{y}$ be such that $\varphi_o(\bar{x}) = F(\bar{x}, \bar{y})$.*

Proof Assume that there exists $\bar{y} \in S(\bar{x})$ such that $\varphi_o(\bar{x}) = F(\bar{x}, \bar{y})$ such that $(\bar{x}, \bar{y})$ is not a local solution or local minimizer of (OBP). Hence, there exists a sequence $(x^k, y^k) \to (\bar{x}, \bar{y})$ with $y^k \in S(x^k)$ such that $F(x^k, y^k) < F(\bar{x}, \bar{y})$. This shows that $\varphi_o(x^k) < \varphi_o(\bar{x})$, and this contradicts the fact that $\bar{x}$ is a local minimizer of (BP_o). This completes the proof.

The next natural question is what about the converse. This means does the local solution of (OBP) corresponds to the local solutions of (BP_o). However, the answer does not seem very straightforward. Note that in case of a global minimum the converse holds while for a local minimum we will need some more conditions.

Proposition 3.2 *Let $(\bar{x}, \bar{y})$ be the global minimizer of (OBP). Then, $\bar{x}$ is a global minimizer of the problem (BP_o).*

Proof On the contrary, let us assume that $\bar{x}$ is not a global minimizer of the problem (BP_o). Then, there exists $\hat{x}$ such that $\varphi_o(\hat{x}) \leq \varphi_o(\bar{x})$. By definition of $\varphi_o(\hat{x})$, there exists $\hat{y} \in S(\hat{x})$ such that

$$\varphi_o(\hat{x}) \leq F(\hat{x}, \hat{y}) < \varphi_o(\bar{x}).$$

Thus again by using the definition of $\varphi_o(\bar{x})$ we have

$$F(\hat{x}, \hat{y}) < \varphi_o(\bar{x}) \leq F(\bar{x}, \bar{y})$$

This contradicts the fact that $(\bar{x}, \bar{y})$ is a global minimizer of (OBP).

Now, we will try to focus our concern on converse when we have a local minimizer of the problem (OBP). In order to do that, we should first begin by viewing the solution set map as a set-valued map which has certain properties. The first property that we will impose is that of local boundedness. We will say that the solution set map $S : \mathbb{R}^n \rightrightarrows \mathbb{R}^m$ is locally bounded at $\bar{x} \in \mathbb{R}^n$ such that there exists a neighbourhood V of $\bar{x}$, such that the set $S(V)$ is a bounded set in $\mathbb{R}^m$. We say that S is locally bounded if it is locally bounded at each $x \in \mathbb{R}^n$. The graph of the solution map is denoted by $gphS$ which is given as

$$gphS = \{(x, y) : y \in S(x)\}.$$

We say that S is graph closed if $gphS$ is a closed set.

Proposition 3.3 *Let us consider that the solution set map S of the lower-level problem is locally bounded and graph closed. If $(\bar{x}, \bar{y})$ be a local minimizer of (OBP), then $\bar{x}$ is a local minimizer of (BP_o), provided that S is single-valued at $\bar{x}$.*

Proof Let us consider that on the contrary $\bar{x}$ is not a local minimizer of φ_o. Hence, there exists a sequence $x^k \to \bar{x}$ such that $\varphi_o(x^k) < \varphi_o(\bar{x})$. Hence, for each k there exists $y^k \in S(x^k)$ such that

$$\varphi_o(x^k) \leq F(x^k, y^k) < \varphi_o(\bar{x}),$$

Now, as S is locally bounded $\{y^k\}$ is a bounded sequence and without loss of generality let us assume that $y^k \to y^*$. Since S has a closed graph, this means that $y^* \in S(\bar{x})$. But since S is single-valued at $\bar{x}$, we conclude that $y^* = \bar{y}$. Thus, $(x^k, y^k) \to (\bar{x}, \bar{y})$. Thus, we have from the previous inequality

$$F(x^k, y^k) < \varphi_o(\bar{x}) \leq F(\bar{x}, \bar{y}).$$

This contradicts the fact that $(\bar{x}, \bar{y})$ is a local minimizer of (OBP).

Note that the problem (OBP) can now be written as

$$\min_{x,y} F(x, y) \quad \text{subject to} \quad (x, y) \in gphS.$$

Though we shall focus on the problems where the lower-level problem is convex in the lower-level variable for each choice of the upper-level variable, the graph of the solution set map S is not a convex set which we will demonstrate through an example below and it would bring forth the fact that the (OBP) is intrinsically a non-convex problem.

Example 3.1 Let the lower-level problem be given as follows

$$\min_{y} f(x, y) = xy, \quad \text{subject to} \quad y \in [0, 1], x \in \mathbb{R}, y \in \mathbb{R}.$$

In this case, we see that $S(x) = \{0\}$, when $x > 0$ and $S(x) = \{1\}$ if $x < 0$, and $S(x) = [0, 1]$ when $x = 0$. One can simply plot the graph to see that it is non-convex.

3.2 Tools from Non-smooth Analysis: A Detour

It is clear that in this article we are going to analyse the non-convex problem (OBP). However, the problem (OBP) is in general non-smooth as seen in the examples in Dempe [6]. Note that this is true even if the data of the problem is linear in both variables and $K(x)$ is polyhedral, the overall problem (OBP) is non-smooth and non-convex. Thus, it is essential that we take a short detour to quickly relevant the definition and properties of some important tools of non-smooth analysis that we shall need in the sequel. The tools that we need are essentially the notion of normal cones and the associated subdifferentials. We will begin our discussion with the regular normal cone or the Frechet normal cone.

Definition 3.1 Let C be a closed set in $\mathbb{R}^n$ and let $\bar{x} \in C$. A vector $v \in \mathbb{R}^n$ is called a regular normal or Frechet normal to C at $\bar{x}$ if

$$\langle v, x - \bar{x} \rangle \leq o(\|x - \bar{x}\|), \quad \forall x \in C,$$

where $\dfrac{o(\|x - \bar{x}\|)}{\|x - \bar{x}\|} \to 0$ as $\|x - \bar{x}\| \to 0$.

The set of all regular normals to C at $\bar{x}$ forms a closed convex cone and is denoted as $\hat{N}_C(\bar{x})$. This definition of a normal is very intuitive since it seems to be an attempt to generalize the notion of the normal to a convex set. It is, however, not difficult to show that when C is convex then $\hat{N}_C(\bar{x})$ coincides with the normal cone of convex analysis which is given as

$$N_C(\bar{x}) = \{v \in \mathbb{R}^n : \langle v, x - \bar{x} \rangle \leq 0 \quad \forall x \in C\}.$$

The regular normal cone suffers from a serious drawback. In many cases, there could be points on the boundary of a closed set where it reduces to the trivial cone containing the zero vector only. This prevents us to use it as a main vehicle in expressing the optimality conditions in the non-convex case. This anomaly can be done away with a limiting process which gives rise to the notion of a limiting normal cone or the basic normal cone.

Definition 3.2 A vector $v \in \mathbb{R}^n$ is called a limiting normal or basic normal vector to the closed set C at $\bar{x}$ if there exists a sequence $v^k \to v$ and $x^k \to \bar{x}$ with $v^k \in \hat{N}_C(\bar{x})$. The collection of the limiting normals to C at $\bar{x}$ forms a closed cone and is denoted as $N_C(x)$.

Note that we have chosen the same symbol to denote the limiting normal cone as we have for the normal cone to a convex set since when C is convex the limiting normal cone coincides with the normal cone for a convex set. The second motivation to do so is that role of the limiting normal cone in non-convex optimization almost parallels the role of the normal cone in convex optimization. However, the normal cone need not be in general convex though it is closed. When $\bar{x}$ is an interior point in C, then $N_C(\bar{x}) = \{0\}$. However, on the boundary of C the limiting normal cone is never trivial. The following example will make the above discussion clear.

Example 3.2 Consider the set $C = epif$ where *epif* denotes the epigraph of the function $f(x) = -|x|$, $x \in \mathbb{R}$. By making a simple drawing, one will clearly see that $\hat{N}_C((0,0)) = \{(0,0)\}$, while

$$N_C((0,0)) = cone\{(1,-1)\} \cup cone\{(-1,-1)\},$$

where *coneA* denotes the cone generated by the set A.

For more details on the regular normal cone and the limiting normal cone, see, for example, Rockafellar and Wets [20] and Mordukhovich [17]. We shall now focus on the notion of the coderivative of a set-valued map. The coderivative of a set-valued map will play a vital role in our discussions of the optimality conditions since we have to deal with the solution set mapping of the lower-level problem.

Let $S : \mathbb{R}^n \rightrightarrows \mathbb{R}^m$ be set-valued map (the reader can just think of the solution set map of the lower-level problem). Assume that $(\bar{x}, \bar{y}) \in gphS$. Then, the *coderivative* of S at $(\bar{x}, \bar{y})$ is given by the set-valued map, $D^*S(\bar{x}|\bar{y}) : \mathbb{R}^m \rightrightarrows \mathbb{R}^n$, which is given as

$$D^*S(\bar{x}|\bar{y})(w) = \{v \in \mathbb{R}^n : (v, -w) \in N_{gphS}(\bar{x}, \bar{y})\}.$$

Keeping in the spirit of the subdifferential of a convex function, the basic subdifferential (also called the limiting subdifferential) of a locally Lipschitz function is given as follows.

Definition 3.3 Let $f : \mathbb{R}^n \to \mathbb{R}$ be a locally Lipschitz function. The basic subdifferential to f at $\bar{x}$ is denoted as $\partial f(\bar{x})$ and is denoted as follows

$$\partial f(\bar{x}) = \{v \in \mathbb{R}^n : (v, -1) \in N_{epif}(\bar{x}, f(\bar{x}))\}.$$

The basic subdifferential is a compact set though need not be convex. However, if f is convex, then the basic subdifferential coincides with the subdifferential of a convex function. If $\bar{x}$ is a local minimum of f at $\bar{x}$, then $0 \in \partial f(\bar{x})$. The basic subdifferential reduces to a singleton set consisting of the gradient vector if the function is continuously differentiable at a given point. The name basic subdifferential can be motivated by the fact that the basic subdifferential is defined by using the basic normal cone. For more details on the basic subdifferentials, see, for example, [20] or [17]. In fact, the convex hull of the basic subdifferential for a locally Lipschitz condition f at x coincides with the Clarke subdifferential $\partial^\circ f(x)$ of f at x. The Clarke subdifferential is a fundamental tool in non-smooth and non-convex analysis. The support function of the Clarke subdifferential at x is denoted by $f^\circ(x, h)$, is called the Clarke directional derivative or more informally as Clarke derivative and is given as

$$f^\circ(x, h) = \limsup_{y \to x, \lambda \downarrow 0} \frac{f(y + \lambda h) - f(y)}{\lambda}$$

It is important to keep in mind that when f is convex the Clarke derivative coincides with the directional derivative of a convex function, i.e.

$$f^\circ(x, h) = f'(x, h) = \lim_{\lambda \downarrow 0} \frac{f(x + \lambda h) - f(x)}{\lambda}$$

For further details on the Clarke subdifferential and Clarke derivative, see, for example, Clarke [3]. For this article, the above information will be enough for us to take the first steps to develop the optimality conditions.

3.3 First Look at Optimality

In what follows we shall focus our attention on the problem (OBP) which we shall henceforth refer to as the *optimistic bilevel programming problem*. The problem (OBP) can be posed as a problem of minimizing a function under a generalized equation constraint. Assume that $f(x, y)$ is continuously differentiable. Then, we can describe the solution set map S as follows

$$S(x) = \{y \in \mathbb{R}^n : 0 \in \nabla_y f(x, y) + N_{K(x)}(y)\}.$$

It is implicit in the above representation that $y \in K(x)$ since otherwise $N_{K(x)}(y)$. We can also for our own convenience describe the following map. Let us introduce the set-valued map $N_K : \mathbb{R}^n \times \mathbb{R}^m \rightrightarrows \mathbb{R}$ which we describe as follows. We say $N_K(x, y) = N_{K(x)}(y)$ if $y \in K(x)$ and $N_K(x, y) = \emptyset$ otherwise. This now allows us to rewrite $S(x)$ as

$$S(x) = \{y \in \mathbb{R}^n : 0 \in \nabla_y f(x, y) + N_K(x, y)\}.$$

Let us recall that the problem (OBP) can be written as

$$\min_{x,y} F(x, y), \quad \text{subject to} \quad (x, y) \in gphS.$$

Let F be a continuously differentiable function. If $(\bar{x}, \bar{y})$ is a local minimizer of the problem (OBP), then the necessary optimality condition is given as

$$0 \in \nabla F(\bar{x}, \bar{y}) + N_{gphS}(\bar{x}, \bar{y}). \tag{3.1}$$

For a proof, see Mordukhovich [17]. Now, the above optimality conditions (3.1) guarantee the existence of a vector $(v, w) \in N_{gphS}(\bar{x}, \bar{y})$ such that

$$0 = \nabla F(\bar{x}, \bar{y}) + (v, w).$$

This tells us that

$$-\nabla_x F(\bar{x}, \bar{y}) = v \quad \text{and} \quad - \nabla_y F(\bar{x}, \bar{y}) = w.$$

The necessary optimality condition (3.1) also shows us that

$$-\nabla_x F(\bar{x}, \bar{y}) \in D^* S(\bar{x}|\bar{y})(\nabla_y F(\bar{x}, \bar{y})).$$

Hence, in order to write the optimality conditions we have the following task laid out for us. We have to estimate $D^* S(\bar{x}|\bar{y})(\nabla_y F(\bar{x}, \bar{y}))$. This will be achieved by using the following lemma which appears as Theorem 1 Levy and Mordukhovich [16].

Lemma 3.1 *Let S be a set-valued map from $\mathbb{R}^n \to \mathbb{R}^m$ given by*

$$S(x) = \{y \in \mathbb{R}^m : 0 \in G(x, y) + M(x, y)\},$$

where $G : \mathbb{R}^n \times \mathbb{R}^m \to \mathbb{R}^d$ is smooth and $M : \mathbb{R}^n \times \mathbb{R}^m \rightrightarrows \mathbb{R}^d$ be a set-valued map with a closed graph.
Let the following constraint qualification holds at $(\bar{x}, \bar{y}) \in gphS$. There is no nonzero $v \in \mathbb{R}^d$ which satisfies

$$0 \in \nabla G(x, y)^T v + D^* M((\bar{x}, \bar{y})| - G(\bar{x}, \bar{y}))(v). \tag{3.2}$$

Then, we have

$$D^* S(\bar{x}|\bar{y})(y^*) \subseteq Q, \tag{3.3}$$

where Q is given by

$$Q = \{x^* \in \mathbb{R}^n : \text{there exists } v^* \in \mathbb{R}^d; (x^*, -y^*) \in \nabla G(x, y)^T v^* +$$
$$D^* M\left((\bar{x}, \bar{y})| - G(\bar{x}, \bar{y})\right)(v^*)\}.$$

In particular if we consider G to be smooth while $M(x, y) = M(y)$ for all x and if $\nabla_x G(\bar{x}, \bar{y})$ is of full rank, then qualification condition (3.2) as mentioned above automatically holds and equality holds in (3.3).

This result can be easily used to derive optimality condition for the problem (OBP). In our case, we have

$$S(x) = \{y \in \mathbb{R}^m : 0 \in \nabla_y f(x, y) + N_{K(x)}(y)\}.$$

Using the representation of the normal cone as a set-valued map, we can again recast $S(x)$ as follows

$$S(x) = \{y \in \mathbb{R}^m : 0 \in \nabla_y f(x, y) + N_K(x, y)\}.$$

One can now easily see that it is a straightforward task to apply the Lemma 3.1 to your setting and derive the following result.

Theorem 3.1 *Let $(\bar{x}, \bar{y})$ be a local minimum of (OBP). Let F be a continuously differentiable function, and let f be a twice continuously differentiable function. Let the following qualification holds at $(\bar{x}, \bar{y})$. There is no nonzero $v \in \mathbb{R}^m$ such that*

$$0 \in \nabla(\nabla_y f(\bar{x}, \bar{y}))^T v + D^* N_K((\bar{x}, \bar{y})| - \nabla_y f(\bar{x}, \bar{y})). \tag{3.4}$$

Then, there exists $v^* \in \mathbb{R}^m$ such that

$$0 \in \nabla F(\bar{x}, \bar{y}) + \nabla(\nabla_y f(\bar{x}, \bar{y}))^T v^* + D^* N_K((\bar{x}, \bar{y})| - \nabla_y f(\bar{x}, \bar{y}))$$

In particular let $K(x)$, the feasible set of the lower-level problem be independent of x, i.e. $K(x) = K$, for all $x \in \mathbb{R}^n$ where K is a closed convex set. Further, assume that the matrix $\nabla_{xy}^2 f(\bar{x}, \bar{y}) = \nabla_x(\nabla_y f(\bar{x}, \bar{y}))$ is of full rank, i.e.

$$rank(\nabla_{xy}^2 f(\bar{x}, \bar{y})) = m.$$

Then, the qualification condition (3.4) mentioned above automatically holds and there exists $v^* \in \mathbb{R}^m$, such that the following conditions hold,

(i) $0 = \nabla_x F(\bar{x}, \bar{y}) + \nabla_{xy}^2 f(\bar{x}, \bar{y})^T v^*$
(ii) $0 \in \nabla_y F(\bar{x}, \bar{y}) + \nabla_{yy}^2 f(\bar{x}, \bar{y})^T v^* + D^* N_K(\bar{y}| - \nabla_y f(\bar{x}, \bar{y}))v^*.$

We must admit that the above optimality conditions are truly of no practical use in bilevel programming and they were not meant to. These optimality conditions are just the first step to understand the complexities involved in the optimality conditions of a bilevel programming problem. The main issue of the optimality condition above is the presence of the coderivative in both the expression of the optimality condition and the qualification condition. The proof of the theorem above is simple, and hence, it is left to the reader. One might, however, argue that since the normal cone map N_K plays a pivotal role in the analysis it is natural to expect that we need to have some kind of derivative for the normal cone map in order to develop the necessary conditions for optimality and the coderivative seems to be a natural choice in this setting. For example, in the second part of Theorem 3.1 if K, for example, is represented by a finite number of convex inequalities, then can we estimate the coderivative of the normal cone map N_K. If we can do that, then we can provide an optimality condition completely in terms of the problem data and possibly escaping the use of the coderivative or represent the coderivative of the normal cone map in terms of the coderivative of a simpler set whose computation is known to us. We would warn the reader at this point that even if we succeed to represent the coderivative of the normal cone map in terms of the coderivative of the simpler set the resulting optimality condition would not be easy to verify. Even if we do not make any progress on the practical use of these optimality conditions, we will nevertheless make some progress on the mathematical aspect of the problem which we believe is encouraging keeping in view the level of difficulty of a bilevel problem. To move towards our goal, we would like to take a detour to peep into the strange world of second-order subdifferentials.

3.4 Second-Order Subdifferentials: Another Detour

In analysis, a second-order derivative is either given through the second-order expansion of the function or viewed as the derivative of the derivative. Our question is how to proceed when the function itself is not differentiable. Mordukhovich and Outrata [18] considered the second point of view and defined the second-order subdifferential as the coderivative of the limiting subdifferential map. In our study of the optimality conditions, we shall need to consider the normal cone map which is defined to be the limiting subdifferential of the indicator function of a closed set. Thus as a first step, we shall concentrate on defining the limiting subdifferential for a proper lower-semicontinuous function.

A function $f : \mathbb{R}^n \to \overline{\mathbb{R}}$, where $\overline{\mathbb{R}} = [-\infty, +\infty]$, is called proper if $f(x) > -\infty$ for all $x \in \mathbb{R}^n$, and there exists $\bar{x} \in \mathbb{R}^n$, such that $f(\bar{x}) < +\infty$. A proper function f is lower-semicontinuous if and only if the epigraph of f is a closed set. This is by now a classical fact to optimizers.

We shall now focus on a key idea required to define the limiting subdifferential in this general setting. This is the notion of the *regular subdifferential*. Note that in order to define the limiting subdifferential for a proper lower-semicontinuous function we follow the approach of Rockafellar and Wets [20]. Let f be a proper lower-semicontinuous function. Then, $v \in \mathbb{R}^n$ is called a regular subgradient of f at

$\bar{x}$ where $f(\bar{x})$ is finite if for all $x \in \mathbb{R}^n$ we have

$$f(x) \geq f(\bar{x}) + \langle v, x - \bar{x} \rangle + o(\|x - \bar{x}\|).$$

If $f(\bar{x}) = +\infty$, then there does not exist a regular subgradient at $\bar{x}$. Thus when $f(\bar{x})$ is finite we call the collection of regular subgradients as the regular subdifferential and is denoted as $\hat{\partial}(\bar{x})$ and if $f(\bar{x}) = +\infty$ then we define $\hat{\partial}f(\bar{x}) = \emptyset$. We would also like to warn the reader that the regular subdifferential can also be empty at a point where f is finite even if the function is locally Lipschitz. Note that the regular subdifferential is always a convex and closed set and coincides with the subdifferential of the convex function if the function f is convex. Let us now proceed to define the notion of a limiting subdifferential for a proper lower-semicontinuous function.

Let f be a proper lower-semicontinuous function, and let $\bar{x}$ be a point where $f(\bar{x})$ is finite. Then, $v \in \mathbb{R}^n$ is said to be a limiting subgradient to f at $\bar{x}$ if there exist sequences $x^k \to \bar{x}$ with $f(x^k) \to f(\bar{x})$ and $v^k \to v$ such that $v^k \in \hat{\partial}f(x^k)$.

The set of all limiting subgradients at $\bar{x}$ constitute what we call the limiting subdifferential of f at $\bar{x}$ which we denote as $\partial_L f(\bar{x})$. As always we shall define $\partial_L f(\bar{x}) = +\infty$ if $f(\bar{x}) = +\infty$. We would also like to mention that in the recent literature on variational analysis the limiting subdifferential is often referred to as the Mordukhovich subdifferential. In contrast to the regular subdifferential, the limiting subdifferential though closed is not a convex set in general and we always have the inclusion $\hat{\partial}f(\bar{x}) \subset \partial_L f(\bar{x})$. Further if f is locally Lipschitz, then the limiting subdifferential is always non-empty, is compact and is related with the celebrated Clarke subdifferential, $\partial^\circ f(\bar{x})$ in the following manner,

$$\partial^\circ f(\bar{x}) = \operatorname{conv}\partial_L f(\bar{x}),$$

where conv denotes the operation of taking the convex hull of a set. For more details on the Clarke subdifferential, a good source to refer to is a recent monograph by Clarke [4]. Moreover, for a proper lower-semicontinuous function we can also have the following representation for the limiting subdifferential given as

$$\partial_L f(\bar{x}) = \{v \in \mathbb{R}^n : (v, -1) \in N_{epif}(\bar{x}, f(\bar{x}))\}.$$

Let us also mention that if $\bar{x}$ is a local minimizer of f over $\mathbb{R}^n$ then we have $0 \in \partial_L f(\bar{x})$. We would like to leave it to the readers to have fun in proving that $\hat{\partial}\delta_C(\bar{x}) = \hat{N}_C(\bar{x})$ and $\partial_L \delta_C(\bar{x}) = N_C(\bar{x})$, where δ_C denotes the indicator function which returns a value zero if $x \in C$ and a takes on the value $+\infty$ if $x \notin C$. For more details on the limiting subdifferential, see Rockafellar and Wets [20], Mordukhovich [17] and Clarke [3].

We are now in a position to introduce the notion of the second-order subdifferential of a proper lower-semicontinuous function. Let $f : \mathbb{R}^n \to \overline{\mathbb{R}}$ be a proper lower-semicontinuous function, and let $\bar{x}$ be a point where $f(\bar{x})$ is finite. Given $\bar{y} \in \partial_L f(\bar{x})$, the second-order subdifferential $\partial_L^2 f(\bar{x}, \bar{y})$ of f at $\bar{x}$ relative to $\bar{y}$ is given as

$$\partial_L^2 f(\bar{x}, \bar{y})(y^*) = D^* \partial_L f(\bar{x}|\bar{y})(y^*)$$

Thus, we can also write

$$\partial_L^2 f(\bar{x}, \bar{y})(y^*) = \{x^* \in \mathbb{R}^n : (x^*, -y^*) \in N_{gph\partial_L f}(\bar{x}, \bar{y})\}.$$

Further if f is C^2 and then setting $\bar{y} = \nabla f(\bar{x})$, we have

$$\partial_L^2 f(\bar{x}, \nabla f(\bar{x}))(y^*) = \{\nabla^2 f(\bar{x})^T y^*\}.$$

The above notion of the second-order subdifferential was first introduced in Mordukhovich and Outrata [18]. In our case in order to compute the coderivative of the normal cone map, we would need to use a calculus rule for computing the second-order subdifferential of a composition map. This calculus rule first appeared in Mordukhovich and Outrata [18] where it appeared in the form of an inclusion and later appeared in Morduhkovich [17]. We shall now present the calculus rule in its latest form. We essentially consider the following composition map $\varphi(x) = (\psi \circ h)(x)$, where $h : \mathbb{R}^n \to \mathbb{R}^m$ and $\psi : \mathbb{R}^m \to \overline{\mathbb{R}}$ is proper and lower-semicontinuous.

Lemma 3.2 *Let φ be finite around $\bar{x}$, let ψ be lower-semicontinuous and proper, h is continuously differentiable around $\bar{x}$, and Jacobian matrix $\nabla h(\bar{x})$ is of full row-rank m. Assume further that ∇h is continuously differentiable around $\bar{x}$; i.e. h is twice differentiable around $\bar{x}$. Let $\bar{y} \in \partial_L \varphi(\bar{x})$ for the composition $\varphi = \psi \circ h$. Let $\partial_L \psi$ be graph closed around $(h(\bar{x}), \bar{v})$, where $\bar{v}$ is the unique vector satisfying*

$$\bar{y} = (\nabla h(\bar{x}))^T \bar{v}, \quad \bar{v} \in \partial_L \psi(h(\bar{x})).$$

Then, we have

$$\partial_L^2 \varphi(\bar{x}, \bar{y})(y^*) = (\nabla^2 \langle \bar{v}, h \rangle(\bar{x})) y^* + (\nabla h(\bar{x}))^T \partial_L^2 \psi(h(\bar{x}), \bar{v})(\nabla h(\bar{x}) y^*).$$

Let us now see how the above result can help us in computing the coderivative of the normal cone map for a convex set K given by the following expression.

$$K = \{x \in \mathbb{R}^n : g_i(x) \leq 0, i = 1, \ldots m\},$$

where each g_i is a convex function. If we write $G(x) = (g_1(x), \ldots g_m(x))$, then we can write

$$K(x) = \{x \in \mathbb{R}^n : G(x) \in -\mathbb{R}_+^m\}.$$

This would allow us to write the following composition

$$\delta_K(x) = \delta_{-\mathbb{R}_+^m} \circ G(x).$$

Observe that we have the following

$$\partial^2 \delta_K(\bar{x}, \bar{y})(y^*) = D^*(\partial_L \delta_K(\bar{x}|\bar{y})(y^*) = D^* N_K(\bar{x}|\bar{y})(y^*).$$

Now using Lemma 3.2 we can conclude that

$$D^* N_K(\bar{x}|\bar{y})(y^*) = (\nabla^2 \langle \bar{v}, G \rangle(\bar{x}))y^* + (\nabla G(\bar{x}))^T D^* N_{\mathbb{R}^m_-}(G(\bar{x})|\bar{v})(\nabla G(\bar{x})y^*),$$

where $\bar{v}$ is the unique solution of the system

$$\bar{y} = \nabla G(\bar{x})^T \bar{v}, \quad \bar{v} \in N_{\mathbb{R}^m_-}(G(\bar{x})).$$

Note that in order to apply Lemma 3.2 we must assume that $\nabla G(\bar{x})$ must be of full row-rank m. Then, one can easily show that

$$\bar{v} = (\nabla G(\bar{x})\nabla G(\bar{x})^T)^{-1}\nabla G(\bar{x})\bar{y}$$

Note that it is not simple to compute the coderivative of normal cone map $N_{\mathbb{R}^m_-}$. See, for example, Proposition 3.7 in Ye [23], and also see Mordukhovich and Outrata [19] or Henrion, Outrata and Surowiec [13].

Now as an application of the discussion in the section let us turn to the optimality conditions that we had presented in the second part of Theorem 3.1. The optimality conditions will now read as follows. Let $(\bar{x}, \bar{y})$ be the solution of the problem (OBP), and assume the hypothesis considered in the second part of Theorem 3.1. Then, we can conclude the existence of $v^* \in \mathbb{R}^m$ such that

(i) $0 = \nabla_x F(\bar{x}, \bar{y}) + \nabla^2_{xy} f(\bar{x}, \bar{y})^T v^*.$
(ii) $0 \in \nabla_y F(\bar{x}, \bar{y}) + \nabla^2_{yy} f(\bar{x}, \bar{y})^T v^* + \nabla^2 \langle \bar{v}, G \rangle(\bar{x})v^* + (\nabla G(\bar{y}))^T D^* N_{\mathbb{R}^m_-}(G(\bar{x})|\bar{v})$
   $(\nabla G(\bar{x})v^*).$

Note that in the special case when K is given as follows

$$K = \{x \in \mathbb{R}^n : Ax \le b\},$$

where A is a $m \times n$ matrix with full row-rank m. Then, it is simple to show that

$$D^* N_K(\bar{x}|\bar{y})(y^*) \subseteq A^T D^* N_{\mathbb{R}^m_-}(A\bar{x} - b|\bar{v})(Ay^*),$$

where $\bar{v} = (AA^T)^{-1}A\bar{y}$. Hence, we have finished our detour and will continue our discussion on the optimality conditions in the next section.

3.5 Value Function Reformulation and Fully Convex Lower-Level Problem

In our discussion till now focus mainly on the problem (OBP) with smooth data. What happens if the lower-level problem has a non-smooth objective. It will then be difficult to develop optimality conditions using the approach we have used till now. Thus, we shall now look into what is called the *value function formulation*. Before we present the value function formulation for the problem (OBP), let us mention the following assumptions that will be valid throughout this section.

(a) $F(x, y)$ is jointly convex in (x, y) and is smooth.
(b) $f(x, y)$ is jointly convex in (x, y) and need not be differentiable everywhere.
(c) Let $K(x) = K$ for all x, and let K be a compact convex set.

Let us define

$$\varphi(x) = \inf_{y \in K} f(x, y)$$

It is clear that φ is a finite-valued convex and continuous function. The value function reformulation of (OBP) is given as the problem (r-OBP),

$$\min F(x, y) \quad \text{subject to} \quad \theta(x, y) = f(x, y) - \varphi(x) \leq 0, y \in K.$$

Note that every local minimizer of (OBP) is a local minimizer of (r-OBP) and vice versa. Our aim would be now to develop the necessary optimality condition for (OBP) through the problem (r-OBP). Before we say anything further it is important to note that θ in the problem (r-OBP) is directionally differentiable. Thus, if $(\bar{x}, \bar{y})$ is a local minimizer of (OBP), then the following system has no solution $(h, w) \in \mathbb{R}^n \times \mathbb{R}^m$.

$$F'(\bar{x}, \bar{y}, h, w) < 0, \quad \theta'(\bar{x}, \bar{y}, h, w) < 0, \quad (h, w) \in T_{\mathbb{R}^n \times K}(\bar{x}, \bar{y}),$$

where $T_{\mathbb{R}^n \times K}(\bar{x}, \bar{y})$ denotes the Bouligand tangent cone to $\mathbb{R}^n \times K$ at $(\bar{x}, \bar{y})$ (see, for example, Bazarra, Sherali and Shetty [1]). Hence, we can conclude that the following system does not have any solution (h, w),

$$F'(\bar{x}, \bar{y}, h, w) < 0, \quad \theta^{\circ}(\bar{x}, \bar{y}, h, w) < 0, \quad (h, w) \in T_{\mathbb{R}^n \times K}(\bar{x}, \bar{y}).$$

Now applying Gordan's Theorem of Alternative (see, e.g. Craven [5]) we conclude that there exists $0 \neq (\lambda_0, \lambda_1) \in \mathbb{R}_+ \times \mathbb{R}_+$ such that

$$\lambda_0 F'(\bar{x}, \bar{y}, h, w) + \lambda_1 \theta^{\circ}(\bar{x}, \bar{y}, h, w) \geq 0, \quad \forall (h, w) \in T_{\mathbb{R}^n \times K}(\bar{x}, \bar{y}).$$

Now by applying the standard support function calculus from convex analysis (see, e.g. Hiriart-Urruty and Lemarechal [14]) we get the following Fritz John type necessary optimality conditions for (OBP), i.e.

(i) $0 \in \lambda_0 \nabla F(\bar{x}, \bar{y}) + \lambda_1 (\partial f(\bar{x}, \bar{y}) - (\partial \varphi(\bar{x}) \times \{0\}) + \{0\} \times N_K(\bar{y})$.
(ii) $0 \neq (\lambda_0, \lambda_1) \in \mathbb{R}_+ \times \mathbb{R}_+$.

In order to show that $\lambda_0 \neq 0$, the most natural thing to do is to impose a qualification basic qualification condition due to Rockafellar and Wets [20] on (r-ORB). The basic constraint qualification is the non-smooth version of the famous Mangasarian–Fromovitz constraint qualification. In our case, the *basic constraint qualification condition* is given said to be satisfied at $(\bar{x}, \bar{y})$ for (OBP) if :

$$\lambda_1 \geq 0, \quad 0 \in \lambda_1 (\partial f(\bar{x}, \bar{y}) - (\partial \varphi(\bar{x}) \times \{0\}) + \{0\} \times N_K(\bar{y}),$$

implies that $\lambda_1 = 0$.
We shall show that this qualification condition always fails for (OBP). Noting the fact that $\partial f(x, y) \subseteq \partial_x f(x, y) \times \partial_y f(x, y)$ we can now write the inclusion condition in the above formulation of the qualification condition in terms of two separate inclusions as given as

$$0 \in \lambda_1 (\partial_x f(x, y) - \partial \varphi(x))$$
$$0 \in \lambda_1 (\partial_y f(x, y) + N_K(y)).$$

Now observe the second inclusion. Since $\bar{y} \in S(\bar{x})$, we conclude that $0 \in \partial_y f(x, y) + N_K(y)$. Thus, we can choose $\lambda_1 > 0$. This immediately violates the qualification condition that we had constructed above. Note that in general the approach to show that the Mangasarian–Fromovitz type conditions fail for the problem (OBP) mainly when K is explicitly defined is usually through the route of replacing the lower-level convex problem by its Karush–Kuhn–Tucker conditions (see, e.g. Dempe [6]) and then showing that the resulting mathematical programming problem under equilibrium constraints (MPEC) does not satisfy the Mangasarian–Fromovitz constraint qualification. Keeping in the view the article by Dempe and Dutta [7], it is better to use the value function approach rather the KKT reformulation approach to show the failure of the constraint qualification. In fact, [7] shows that the problem (OBP) and its KKT reformulation are not in general equivalent unless some additional conditions are met.
The question that we face now is that is there any additional condition which allows us to show that $\lambda_0 > 0$ in the above optimality conditions. One such condition is the notion of partial calmness given by Ye and Zhu [21] which we now present below.

Definition 3.4 Consider the perturbed version of the (r-OBP) problem given as

$$\min F(x, y), \quad \text{subject to} \quad f(x, y) - \varphi(x) + u = 0, y \in K, u \in \mathbb{R}.$$

We will say that (OBP) is partially calm at $(\bar{x}, \bar{y})$ if there is a constant $M > 0$ and an open neighbourhood of $(\bar{x}, \bar{y}, 0)$ such that for each feasible (x, y, u) of the perturbed version of the (r-OBP) problem we have

$$F(x, y) - F(\bar{x}, \bar{y}) + M |u| \geq 0.$$

Ye and Zhu [21] provided the following more usable charecterization of the partial calmness condition.

Lemma 3.3 *Let $(\bar{x}, \bar{y})$ be a local minimizer of (OBP). Then, (OBP) is partially calm at $(\bar{x}, \bar{y})$ if there exists $\lambda > 0$ such that $(\bar{x}, \bar{y})$ is also a local minimizer of the problem*

$$\min(F(x, y) + \lambda(f(x, y) - \varphi(x))), \quad subject\ to \quad y \in K.$$

We would like to note that the notion partial calmness in some sense allows a penalization of the problem (r-OBP). Moreover, we would like to state we are doing the discussion on partial calmness in the framework of the assumption that $K(x) = K$ for all $x \in \mathbb{R}^n$. Further, for a given $\bar{x}$ if all the solutions $\bar{y}$ of the lower-level problem is are uniformly weak sharp then it was shown in [21] that (OBP) is partially calm at $(\bar{x}, \bar{y})$. The notion of uniformly weak sharp minima was introduced in Ye and Zhu [21]. We shall now provide below a brief description of the notion of weak sharp minimum as described in [12].
Consider the lower-level problem of (OBP), where $K(x) = K$ for all x where K is a closed convex set, i.e.

$$\min_{y}\{f(x, y) : y \in K\}$$

We say that the above problem has a *uniformly weak sharp minimum* around one of the solutions $\bar{y} \in C$ computed at $x = \bar{x}$ if there exists $\delta > 0$ and $\alpha > 0$, such that

$$f(x, y) - \varphi(x) \geq \alpha d(y, S(x)),$$

for all x such that $\|x - \bar{x}\| < \delta$ and for all $y \in C$ such that $\|y - \bar{y}\| < \delta$. Here, as before φ represents the optimal value function and $d(y, S(x))$ represents the distance of y from $S(x)$. For details on the notion of weak sharp minimum, see, for example, Burke and Ferris [2].
Let us now illustrate the notion of partial calmness by presenting some examples based on those studied in Dempe, Dutta and Mordukhovich [8]. Consider a problem where the upper-level objective is given as

$$F(x, y) = \frac{(x - 1)^2}{2} + \frac{y^2}{2}, \quad (x, y) \in \mathbb{R}^2.$$

The lower-level problem is given as

$$\min_{y}(\frac{x^2}{2} + \frac{y^2}{2}), \quad subject\ to \quad y \in K = [0, 1]$$

Note that this problem is fully convex and has a compact lower-level feasible set. Here, $S(x) = \{0\}$ and $\varphi(x) = \frac{x^2}{2}$. Thus, $(\bar{x}, \bar{y}) = (1, 0)$ is the only solution of the bilevel problem. Note that

$$f(x, y) - \varphi(x) = \frac{y^2}{2}$$

The penalized problem is thus given for any $\lambda > 0$ as

$$\min \frac{(x-1)^2}{2} + \frac{y^2}{2} + \lambda\frac{y^2}{2}, \quad \text{subject to} \quad y \in K = [0, 1].$$

It is clear that $(\bar{x}, \bar{y}) = (1, 0)$ is the unique solution of the penalized problem for any $\lambda > 0$. Thus, using Lemma 3.3 we conclude that the given problem is partially calm at (OBP).

However, by slightly tweaking the above upper-level objective function and making the lower-level problem unconstrained, we will see that partial calmness can fail. Let the upper-level objective function is now given as

$$F(x, y) = \frac{(x-1)^2}{2} + \frac{(y-1)^2}{2}, \quad (x, y) \in \mathbb{R}^2.$$

The lower-level objective function remains the same at the previous example but the constraints are removed that is $K = \mathbb{R}$. The penalized problem in this case is given as

$$\min \frac{(x-1)^2}{2} + \frac{(y-1)^2}{2} + \lambda\frac{y^2}{2},$$

As before, the point $(\bar{x}, \bar{y}) = (1, 0)$ is the unique solution of the bilevel problem but it is not a solution of the penalized problem for any $\lambda > 0$. Note that for any $\lambda > 0$ the minimizer for the penalized problem is given as $(1, \frac{1}{1+\lambda}) \neq (1, 0)$. Thus, in this case the problem fails to be partially calm. We will now focus on fully convex bilevel problem whose lower-level objective is fully convex but not differentiable everywhere. Consider the bilevel problem with upper-level objective given as

$$F(x, y) = \frac{x^2}{2} + \frac{y^2}{2}, \quad (x, y) \in \mathbb{R}^2.$$

The lower-level problem in this case is given as

$$\min(|x| + |y|), \quad \text{subject to} \quad y \in K = [-1, +1].$$

It is not much difficult to show that the bilevel problem is partially calm at the solution $(0, 0)$.

Note that this penalization property associated with the partial calmness condition immediately shows that we can take $\lambda_0 > 0$. Thus, to bring our discussion to an end we shall now see that we need to estimate the subdifferential of φ in terms of the data of the lower-level problem in (OBP). In order to do such an estimation, we need to impose certain continuity assumptions on the solution set map S. The most useful such assumption is that of the lower-semicontinuity of the map S. However, this assumption is not a natural one in our setting. Another assumption which will work and can hold in our setting is that of inner-semicompactness. Under our assumptions in this section, we say that S is inner-semicompact at $\bar{x}$ if for every sequence $x^k \to \bar{x}$ there is a sequence $y^k \in S(x^k)$ that contains a convergent subsequence. In fact, if S is locally bounded, then it is inner-semicompact. Using the results in Mordukhovich [17], Chap. 3, and using the inner-semicompactness assumption on the map S, we can show that

$$\partial \varphi(\bar{x}) \subset \bigcup_{\bar{y} \in S(\bar{x})} \{x^* \in \mathbb{R}^n : x^* \in \partial_x f(\bar{x}, \bar{y})\}. \tag{3.5}$$

See the appendix at the end of the article to see how one can obtain the estimate (3.5). This allows us to conclude our discussion in this section by stating the following theorem and followed by some examples about its applicability.

Theorem 3.2 *Let us consider the problem (OBP), and let us assume that $(\bar{x}, \bar{y})$ be a local minimizer. The problem data is fully convex, and K is a compact convex set. Let us assume that S is inner-semicompact at $\bar{x}$ and the problem (OBP) is partially calm at $(\bar{x}, \bar{y})$. Then, there exists $\bar{y} \in S(\bar{x})$ and $\lambda_1 > 0$ such that*

(i) $0 \in \nabla_x F(\bar{x}, \bar{y}) + \lambda_1 \partial_x f(\bar{x}, \bar{y}) - \lambda_1 \partial_x f(\bar{x}, \bar{y})$.
(ii) $0 \in \nabla_y F(\bar{x}, \bar{y}) + \lambda_1 \partial_y f(\bar{x}, \bar{y}) + N_K(\bar{y})$.

A very important point is that when the data of the problem is fully linear then Ye and Zhu [21] proved that partial calmness automatically holds. Thus, for Theorem 3.2 to be applicable one must assume that the solution map associated with a fully linear problem is inner-semicompact. The surprising is fact that it might not be so in general. This can be shown using Theorem 3.2. Let us assume that $F(x, y) = \langle a, x \rangle + \langle b, y \rangle$, where $0 \neq a \in \mathbb{R}^n$ and $b \in \mathbb{R}^m$. Also, assume that $f(x, y) = \langle c, x \rangle + \langle d, y \rangle$. In this case, we have $\partial_x f(x, y) = \{c\}$ and $\nabla_x F(x, y) = a$ for all (x, y). Let us assume that the solution set map at the solution S is inner-semicompact at $\bar{x}$ where $(\bar{x}, \bar{y})$ is the solution of (OBP) with fully linear data. Then, we can apply (i) in Theorem 3.2 to conclude that $a = 0$ which will contradict the hypothesis. This shows that in this case S is not inner-semicompact at $\bar{x}$. We now show below an example of the problem (OBP) with fully convex (but nonlinear) data where all the assumptions of Theorem 3.2 are satisfied.

Example 3.3 Let us consider the problem (OBP) where

$$F(x, y) = x^2 + y^2,$$

and the lower-level objective is given as $f(x, y) = y^2, y \leq 0, f(x, y) = 0, 0 \leq y \leq 1$ and $f(x, y) = (y - 1)^2, y \geq 1$. Further let $K = [0, \frac{1}{2}]$. Then, it is clear that for all x we have $S(x) = [0, \frac{1}{2}]$. Note that S is clearly inner-semicompact. Note that $(\bar{x}, \bar{y}) = (0, 0)$ is a solution of the (OBP) with the given data in this example. Note that

$\varphi(x) = 0$. Note that it is simple to prove that for any $\lambda > 0$ the point $(0, 0)$ also solves the problem

$$\min_{y \in [0, \frac{1}{2}]} F(x, y) + \lambda(f(x, y) - \varphi(x)).$$

Thus, the problem (OBP) satisfies the partial calmness condition at $(0, 0)$ and it is simple to see that the conditions (i) and (ii) in Theorem 3.2 are satisfied at $(0, 0)$.

3.6 Conclusions

The article is based on the lectures delivered at the CIMPA school held in New Delhi in 2013. These notes aim to just open the door to the challenging area of developing optimality conditions for an optimistic bilevel programming problem. In these notes, we have highlighted two approaches. Through one approach, we develop necessary optimality conditions where second-order derivative information is critical. However, as we saw the use of partial calmness property has allowed us to use only first-order information. However, we see that it is not so easy to check the partial calmness condition. For example, Henrion and Suwoviec [12] have used qualifications based on the standard calmness property but they also had to use second-order derivative information. For more recent work on the necessary conditions for optimistic bilevel programming and their applications, see, for example, Dempe and Zemkoho [9] and Lin, Xu and Ye [25].

Acknowledgements I am grateful to the organizers of the CIMPA school on Generalized Nash Equilibrium, Bilevel Programming and MPEC problems held in Delhi from 25 November to 6 December 2013 for giving me a chance to speak at the workshop which finally led to this article. I would also like to thank Didier Aussel for his thoughtful discussions on the article and his help in constructing Example 3.3. I am also indebted to the anonymous referee for the constructive suggestion which improved the presentation of this article.

Appendix

We are now going to demonstrate how to obtain the estimation of $\partial \varphi(\bar{x})$ as given in (3.5). We shall begin with a general setting of a parametric optimization problem given as

$$\min_{y} f(x, y) \quad \text{subject to} \quad y \in K(x).$$

where $f(x, y)$ is say an extended-valued proper lower-semicontinuous map function and further K is set-valued map with closed convex and non-empty values for each x.

Further, assume that S is an inner-semicompact map. Our question is that can we estimate $\partial\varphi(\bar{x})$ where φ is given as

$$\varphi(x) = \inf_y \{f(x, y) : y \in K(x)\}.$$

In order to do so, we need to consider the notion of the *asymptotic subdifferential* or *singular subdifferential* of a lower-semicontinuous function. Note that if h is a proper lower-semicontinuous function then we define the asymptotic subdifferential at $\bar{x}$ as

$$\partial_L^\infty h(\bar{x}) = \{v \in \mathbb{R}^n : (v, 0) \in N_{epif}(\bar{x}, h(\bar{x}))\}.$$

In the context of the given parametric optimization problem in this section, let us assume that the following qualification condition holds at $(\bar{x}, \bar{y}) \in gphS$. The condition says that

$$\partial_L^\infty f(\bar{x}, \bar{y}) \cap (-N_{gphK}(\bar{x}, \bar{y})) = \{(0, 0)\}. \qquad (3.6)$$

If this qualification condition holds, then Mordukhovich [17] shows us that

$$\partial_L\varphi(\bar{x}) \subseteq \bigcup_{\bar{y}\in S(\bar{x})} \{x^* + D^*K(\bar{x}|\bar{y})(y^*) : (x^*, y^*) \in \partial_L f(\bar{x}, \bar{y})\}. \qquad (3.7)$$

Note that since we consider f to be finite-valued and jointly convex we conclude that f is locally Lipschitz and hence we know from Mordukhovich that

$$\partial_L^\infty f(\bar{x}, \bar{y}) = \{(0, 0)\}.$$

Hence, the qualification condition given in (3.6) is automatically holding. Further note that as $K(x) = K$ we have for any $y^* \in \mathbb{R}^m$.

$$D^*K(\bar{x}|\bar{y})(y^*) = \{0\}.$$

This shows that in our setting of a fully convex problem with a compact and convex K we have

$$\partial\varphi(\bar{x}) \subseteq \bigcup_{\bar{y}\in S(\bar{x})} \{x^* \in \mathbb{R}^n : x^* \in \partial_x f(\bar{x}, \bar{y})\}.$$

Exercises

1. Write down the proof of Theorem 3.1 in detail.
2. Let us now assume that in the problem (OBP) the feasible set of the lower-level problem is independent of x, i.e. $K(x) = K$. Further, let us assume that $K = \{x \in \mathbb{R}^n : g_i(x) \le 0, i = 1, \ldots p\}$, where each g_i is a convex function. With this additional information how can you write down the problem (OBP) in terms

of the optimality conditions of the lower-level problem. Note that the constraints functions g_i, $i = 1, \ldots, p$ must now appear in the model. Assuming that each local minimum of the problem (OBP) is also a local minimum of the problem where the lower-level is represented in terms of optimality conditions we urge the reader to now write down the optimality condition for the bilevel problem. Choose appropriate assumptions.

3. This is possibly an open question still to best of our knowledge. Can we replace the notion of partial calmness by some more tractable condition. One can of course see [12] but again calmness conditions are not so easy to check. It would be indeed important to have a better alternative to partial calmness.
4. Write down the proof of Theorem 3.2 in detail.

References

1. Bazaraa, M.S., Sherali, H.D., Shetty, C.M.: Nonlinear programming. Theory and algorithms, 3rd edn. Wiley-Interscience, Wiley, Hoboken (2006)
2. Burke, J.V., Ferris, M.C.V.: Weak sharp minima in mathematical programming. SIAM J. Control Optim. **31**, 1340–1359 (1993)
3. Clarke, F.: Optimization and Nonsmooth Analysis. Wiley-Interscience, Hoboken (1983)
4. Clarke, F.: Functional Analysis, Calculus of Variations and Optimal Control. Graduate Texts in Mathematics, vol. 264. Springer, London, 2013
5. Craven, B.D.: Mathematical Programming and Control Theory. Chapman and Hall, London (1978)
6. Dempe, S.: Foundations of Bilevel Programing (Springer). Kluwer Academic Publishers, The Netherlands (2003)
7. Dempe, S., Dutta, J.: Is bilevel programming a special case of a mathematical program with complementarity constraints? Math. Program. Ser. A **131**, 37–48 (2012)
8. Dempe, S., Dutta, J., Mordukhovich, B.S.: New necessary optimality conditions in optimistic bilevel programming. Optimization **56**, 577–604 (2007)
9. Dempe, S., Zemkoho, A.B.: The bilevel programming problem: reformulations, constraint qualifications and optimality conditions. Math. Program. Ser. A **138**(1–2), 447–473 (2013)
10. Dempe, S., Mordukhovich, B., Zemkoho, A.B.: Necessary optimality conditions in pessimistic bilevel programming. Optimization **63**(4), 505–533 (2014)
11. Dempe, S., Dutta, J.: Bilevel programming with convex lower level problems. Optimization with multivalued mappings, pp. 51–71, Springer Optim. Appl., 2, Springer, New York, (2006)
12. Henrion, R., Surowiec, T.: On calmness conditions in convex bilevel programming. Appl. Anal. **90**, 951–970 (2011)
13. Henrion, R., Outrata, J., Surowiec, T.: On the co-derivative of normal cone mappings to inequality systems. Nonlinear Anal. **71**, 1213–1226 (2009)
14. Hiriart-Urruty, J.B., Lemarechal, C.: Fundemantals of Convex Analysis. Springer, Berlin (2003)
15. Jahn, J.: Vector Optimization. Applications, and Extensions. Springer-Verlag, Berlin, Theory (2004)
16. Mordukhovich, B.S., Levy, A.B.: Coderivatives in parametric optimization. Math. Program. Ser. A **99**(2), 311–327 (2004)
17. Mordukhovich, B.S.: Variational analysis and generalized differentiation Vol 1 Grundlehren der Mathematischen Wissenschaften [Fundamental Principles of Mathematical Sciences], 331. Springer-Verlag, Berlin (2006)
18. Mordukhovich, B.S., Outrata, J.V.: On second-order subdifferentials and their applications. SIAM J. Optim. **12**, 139–169 (2001)

19. Mordukhovich, B.S., Outrata, J.V.: Coderivative analysis of quasi-variational inequalities with applications to stability and optimization. SIAM J. Optim. **18**, 389–412 (2007)
20. Rockafellar, R.T., Wets, R.J.B.: Variational Analysis. Springer, Berlin (1997)
21. Ye, J.J., Zhu, D.: Optimality conditions for bilevel programming problems. Optimization **33**, 9–27 (1995)
22. Ye, J.J.: Nondifferentiable multiplier rules for optimization and bilevel optimization problems. SIAM J. Optim. **15**, 252–274 (2004)
23. Ye, J.J.: Constraint qualifications and necessary optimality conditions for optimization problems with variational inequality constraints. SIAM J. Optim. **10**, 943–962 (2000)
24. Ye, J.J., Zhu, D.: New necessary optimality conditions for bilevel programs by combining the MPEC and value function approaches. SIAM J. Optim. **20**, 1885–1905 (2010)
25. Lin, G.H., Xu, M., Ye, J.J.: On solving simple bilevel programs with a nonconvex lower level program. Math. Program. Ser. A **144**(1–2), 277–305 (2014)

Chapter 4
Mechanism Design and Auctions for Electricity Networks

Benjamin Heymann and Alejandro Jofré

Abstract In this chapter, we present some key aspects of wholesale electricity markets modeling and more specifically focus our attention on auctions and mechanism design. Some of the results stemming from these models are the computation of an optimal allocation for the Independent System Operator, the study of equilibria (existence and uniqueness in particular) and the design of mechanisms to increase the social surplus. More generally, this field of research provides clues to discuss how wholesale electricity market should be regulated. We begin with a general introduction and then present some results we obtained recently. We also briefly discuss some ongoing related research. As an illustrative example, a section is devoted to the computation of the Independent System Operator response function for a symmetric binodal setting with piece-wise linear production cost functions.

4.1 Introduction

Economists, engineers, and mathematicians have given a lot of attention to electricity markets since the beginning of the liberalization era in the 1980s. We present recent results and ongoing research about wholesale electricity markets and, in particular, the optimal design of such market. The field of market design studies the effects of market rules on economic functioning such as oligopoly behavior, vertical integration, market power, pricing, externalities. The number of recent Nobel Prize laureates with contributions in this field demonstrates its impact on economic thinking. In this chapter, we focus on recent works [1–4] as well as on some ongoing research by the authors. In the following sections, we assume that we are in a mandatory pool market; i.e., the agents will satisfy their engagements.

B. Heymann (✉)
Ecole polytechnique, Palaiseau, France
e-mail: benjamin.heymann@polytechnique.edu

A. Jofré
CMM, Santiago, Chile
e-mail: ajofre@dim.uchile.cl

Diversity in electrical markets models can be explained by market specificity. This specificity stems from its economics, industrial and geographical setting, its dependency on the regulatory environment, timescales, and the complex physical properties of electrical networks (Kirchhoff laws, for instance) as well as the entities that compose it—producers, consumers, Independent System Operator (ISO), and networks.

Key modeling decisions concern the agents' preferences, the uncertainties on the energy sources and demands, information representation, production capacities, and the physics of the system. In particular, one has to specify the structure used to represent the bidding strategies of the producers. Since the physics of an electrical network is a difficult problem too, it is usually simplified.

There are (also) classic questions which accompany any modeling attempt, such as the mathematical well posedness of the problem, and the existence, uniqueness, and tractability of the equilibria as well as their properties. One might also ask about the existence of efficient algorithms for calculating those equilibria. We point out that models for wholesale electricity markets are often general enough to be relevant for other economic settings.

In our setting, the production allocation plan is the result of an auction. The producers communicate their selling prices to a central agent, and then the central agent minimizes the total cost while satisfying the demand. In our model, we most of the time take into account the geography of the network (i.e., production and consumption are not colocalized at one point) as well as the losses due to the electricity transportation. Figure 4.1 presents a simple example of a network with four nodes.

We first present the general setting of the model in Sect. 4.2. Section 4.3 is a short review of some recent relevant work in the field. We give a quantitative formulation of the problem in Sect. 4.4. We will discuss the main results in Sect. 4.5. In Sect. 4.6, we develop the example of two producers setting with piecewise linear cost functions. We conclude in Sect. 4.7.

4.2 Setting

This section is a qualitative description of the market settings encountered in the literature. We try to be as general as possible, whereas in Sect. 4.3 we focus on the frameworks Escobar, Figueroa, Heymann and Jofré study in [1–4]. The main market model components are the agents, the demand, the network, the regulation, and the structure of information (since some uncertainty is usually part of the model). Different types of equilibria could be considered. An example of a network is proposed in Fig. 4.1.

The agents are divided among those who produce electricity (producers) and those who consume it (usually aggregated into an ISO). In our setting, both producers and consumers are macroscopic, but for the sake of completeness, we note that some models use a continuum of microscopic producers. Such models correspond to situations where no producer can have any unilateral impact on the market.

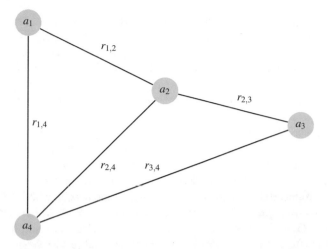

Fig. 4.1 An example of a wholesale electricity market network. On the demand side, at each node a_i there is a demand (load) d_i that has to be fulfilled. On the supply side, electricity can be produced at some nodes by some independent agents. The production cost function associates the quantity produced by an agent and the corresponding economic cost (it is specific to the agent). For modeling but also technical and practical reasons, those production cost functions are often approximated with functions in simple functional sets (linear, quadratic, piecewise linear, etc.). Electricity can be sent from one node to another through the edges of the network, but there is a price for that (for instance, a loss proportional to the square of the quantity sent, due to resistivity). The Independent System Operator (ISO) is as its name indicates a central operator that has to allocate the production so that supply meet demand while minimizing a criterion (usually the total price). To produce this allocation, the ISO needs to know the price he will have to pay for each allocation, so the producers specify their bid functions that are usually in the same simple aforementioned functional set. Since the ISO has no way to know the real production cost function of the producers, it is in their interest to game the system. We point out that stochasticity may occur on the demand. Moreover, the network aspect of the market is of primary importance, as it is responsible for the market power of the agent

Producers incur production costs when they supply electricity to the market. These costs depend on the quantity of electricity they each produce individually. The relation between the quantity a producer supplies and its production cost is encoded in his production cost function. To sell electricity, producers quote a price to the market. We consider two structures to model the way a producer specifies a selling price, which we discuss later in the chapter: the bid function and the supply function. A bid function maps any quantity of electricity to the price a producer asks to supply such a quantity. A supply function maps any price to the quantity a producer is ready to supply at such price. The objective of each producer is to maximize his individual profit (or his average profit if the model contains any form of randomness). We consider only noncooperative settings: producers are competing against each other.

We assume dispatch decisions to be centralized: a unique agent, the Independent System Operator (ISO), aggregates the demand side. We justify this aggregation by the regulatory environment and the market organization. The ISO receives bids from producers and has to supply the local electricity demand where it is needed by buying the electricity at the quoted prices (pay as bid market). Generally, the demand is inelastic (in the literature we are considering), but it could be either deterministic or stochastic.

Usually, electricity is seen as a divisible commodity. Nonetheless, in our model we can also see it as a geographically differentiated product. Unless production facilities and consumers are colocalized, productions and demands are dispatched on a network. The network contains nodes and edges. Each node is a place where electricity is produced or consumed (or both) and as such is characterized by its local demand and its local producers. Each edge is a place through which electricity can be sent. The ISO has the possibility to send any nodal electricity surplus where it is needed through the network cables but those cables are subject to physical limitations such as capacity constraints and online losses (due to Joule effects). The geographical differentiation comes from the fact that it is the ISO which incurs those losses.

In our model, we envision different kinds of optimization problems: the standard ISO problem, the agent's profit maximization problem, and the mechanism design problem. The first one consists simply in finding the minimal cost production plan. For this problem, the ISO (or the principal, if we wish to use mechanism design terminology) receives bids from the different producers and knows the demand (deterministic or stochastic) at each node. He then has to supply this demand at each node for the cheapest total cost. This optimization is subject to the network's physical limitations and is parametrized by the demand at each node and the producers' bids. We call the function that maps these parameters with the solution of the corresponding problem *principal response*.

The second problem stems from the agents' perspectives. Knowing the principal response function, their own production cost functions and possessing some common knowledge on their fellow producers, they optimize their bids to maximize their individual profits. This problem raises questions about best response strategies and Nash equilibria.

As alluded previously, in mechanism design, there is an agent and a principal. In our model, we assign the role of the principal to the ISO. The mechanism design problem reverts the role of the principal and the agents: The principal builds his response function knowing that the agents will then maximize their profits. By offering the right incentives, he leads the agents. The mechanism design problem can be formulated by considering that the principal gives a new response function that is not the optimal solution to the first problem. Indeed, instead of waiting for the bids to order the production, the ISO defines in a contract a response function that he will respect in the future. This contract depends on the (future) bids of each producer and the demand at each node. This occurs because otherwise there would not exist incentives to convince the agents to tell the truth about their production costs. They would optimize their own benefits based on the information they have. We have shown that in some very general setting, it is possible for the principal to formulate the response

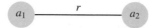

Fig. 4.2 In [2, 4], the authors consider a binodal market with quadratic line losses. The demand is the same at both nodes. The production cost functions are linear. There are no network and capacity constraints. A very intuitive justification of the market power induced by the line losses is given in [2]. Indeed, in a symmetric perfect information setting with linear production cost function of slope c, the equilibrium strategy of both producers is to bid $\frac{c}{1-2dr} > c$

function in the contract so that the producers are incentivized to reveal their true types (i.e., real production costs). Put differently, the principal can design a contract so that for each producer, it is optimal to reveal his true production costs function. To do so, the principal has to pay (virtually, through the payment function defined in the contract) an information rent to the producers, but his total cost is smaller than in the previous setting.

In general, the principal does not know the real production cost of the producers. This is the reason why producers can bid higher than their production cost. The information the principal has about producers' costs is modeled by a probability distribution. The less the producer knows about production costs, the higher the information rent (Fig. 4.2).

4.3   Literature

Several approaches have been proposed to answer the questions raised in the previous section. In [5], Klemperer and Meyer show that uncertainty reduces the quantity of supply function Nash equilibria. The firms bid their supply functions before demand is revealed. The existence of a Nash equilibrium is shown for a symmetric oligopoly. In [6], Anderson and Philpott show how to construct optimal time-dependent supply functions in electricity market settings where demand and competing generators' behaviors are unknown by introducing a market distribution function. The gaming aspect of the situation is reduced by arguing that competitors do not react to the producer bids. The problem is formulated as an optimal control problem. In [7], the authors study asymmetric competition and propose a numerical solver based on GAMS to compute the optimal strategies. They compare the algorithm with the ODE method. In [8], Anderson gives a proof of existence of a pure Nash equilibrium under some technical assumptions when the network is reduced to a single point. He also gives sufficient conditions for uniqueness. Optimal auction design was introduced by Myerson in his 1981 seminal article [9]. Laffont and Martimort wrote in [10] an introduction to mechanism design in a general setting. These authors expose important concepts such as the revelation principle, adverse selection, participation constraints, and information rent. The book does not consider interactions on a network—which is the specificity of the wholesale electricity market. Bi-level approaches with quadratic production cost functions are proposed in [11, 12] to study the ISO response functions and the Nash equilibria. For a reference to complementarity modeling, the reader can consult [13]. Escobar and Jofré show in [3] that in a random environment Walrasian and noncooperative equilibria exist (for the noncooperative equilibrium, the distribution needs to be atom-less) in this setting, the demand is elastic, and the ISO maximizes the sum of the utility functions. Utility functions and cost functions are general.

4.4 Quantitative Formulations

We briefly present in this section some questions of interest concerning models that fit into the general setting described in Sect. 4.2. Those questions were partially addressed in recent works by the authors.

4.4.1 Generality

We will generically use the notations i to refer to a node or its corresponding producer and q_i to refer to the quantity this agent produces. The nodes are connected by edges, and we denote by $h_{i,i'}$ the quantity of electricity that is sent from node i to node i'. The market network is not necessarily complete. We call d_i the demand at node i. Each producer quotes a bid denoted by b_i to the principal. This bid is a function of the quantity q_i. Each producer also informs the principal of the maximum quantity $\bar{q}_i$ he can produce. In general, the allocation problem is subject to network constraints; i.e., the vector h of components $h_{i,i'}$ has to be in a set H. For example, the set H could be made of all vectors h such that $h_{i,i'} \leq h_{i,i'}^{max}$, which means that one cannot send an arbitrary big amount of electricity through the network.

4.4.2 The Standard Allocation Problem

The principal receives bids from the agents and allocates the production so that:

- the allocation minimizes the total cost;
- the allocation respects the network and capacity constraints;
- supply is greater than demand at any node.

The last point corresponds to the nodal constraints. The supply at a given node i is the sum of the local production q_i and the importations from neighboring nodes $\sum_{i'} h_{i',i}$. To this, we need to subtract the exportations to neighboring nodes $\sum_{i'} h_{i,i'}$ and the line losses. If we send a quantity h through an edge $\{i, i'\}$, we denote by $L_{i,i'}(h)$ the corresponding loss. We will count half of this loss at node i and the other half at node i'. We end up with the following nodal constraint:

$$q_i + \sum_{i'} [h_{i',i} - h_{i,i'}] - \sum_{i'} \frac{L_{i',i}(h_{i',i}) + L_{i,i'}(h_{i,i'})}{2} \geq d_i, \qquad (4.1)$$

where the summations are performed over the nodes adjacent to i. All this being said, the generic allocation problem writes

$$\begin{aligned}
\underset{q,h}{\text{minimize}} \quad & \sum_i b_i(q_i) \\
\text{subject to} \quad & q_i + \sum_{i'} [h_{i',i} - h_{i,i'}] - \sum_i \frac{L_{i,i'}(h_{i',i}) + L_{i,i'}(h_{i,i'})}{2} \geq d_i \\
& q_i \leq \bar{q}_i \\
& q_i \geq 0 \\
& h \in H \\
& h_{i,j} \geq 0.
\end{aligned} \qquad (4.2)$$

We point out that if the bidding and the loss functions are convex functions and H is a convex set, then the problem is convex. For instance, one can take the bid functions linear, the loss functions quadratic, and H equal to $\mathbb{R}^+$. Observe that for a convex problem, if L is strictly convex then at optimality $h_{i',i}h_{i,i'} = 0$. Note that the bid functions b_i and the demand vector d can be seen as parameters of the optimization problem. We could make the solution of this problem stochastic by adding a dependency of d to a random variable ω. This would not change the solution of the problem from the operator perspective, but it would change the market setting for the agents. **What is the solution of this deterministic allocation problem? What are the analytical properties of this solution? How can we compute it?**

4.4.3 The Agent Problem

The objective of each producer is to maximize his profit. Note that by solving the principal allocation problem, we have the response function of the ISO to the agents' bids. It is stochastic if the demand is stochastic. We can map each bidding profile of the agents with the expected profit of each agent. By competing against each other, the agents are playing a game. In addition, producer i does not know the production cost functions of his fellow agents. As a result, we are in an imperfect information setting. We assume that for each agent i there is a probability distribution f_i over a set of potential production cost functions. The probability distribution f_i represents the (common) information the other agents have about agent i. We assume those probability distributions to be independent. The profit of an agent of type (i.e., production cost function) c_i that bids b_i (that associates any production level with the price he asks or such production level) is given by

$$\pi_i(c_i, b_i) = \int_{C_{-i}} [b_i(q_i(b_i, b_{-i}(c_{-i})) - c_i(q_i(b_i, b_{-i}(c_{-i})))] f_{-i}(c_{-i}) dc_{-i}, \qquad (4.3)$$

where the integral is performed over the types of the other agents. In this expression, $q_i(b_i, b_{-i})$ corresponds to the production level of producer i in the ISO allocation plan when the bids are (b_i, b_{-i}). The production cost function c_i associates the production level q_i and the corresponding true cost for producer i: $c_i(q_i)$. The function $b_{-i}(c_{-i})$ is the vector of bidding functions of the other producers when their types are the vector c_{-i}. Then, the maximized profit is

$$\overline{\pi}_i(c) = \max_{b_i} \int_{C_{-i}} [(b_i(q_i(b_i, b_{-i}(c_{-i})) - c(q_i(b_i, b_{-i}(c_{-i})))] f_{-i}(c_{-i}) dc_{-i}. \qquad (4.4)$$

So for each agent the best response strategy to the other agents is the solution of an optimization problem on the set of maps from the types (i.e., production cost functions) c to the bids b. Usually, the production cost functions will be characterized by a vector of $\mathbb{R}^n$. In this case, this is an optimization over the functions from $\mathbb{R}^n$ to $\mathbb{R}^n$. Observe that this setting corresponds to a (Bayesian) Bertrand game. Of course, it is natural to ask about the Nash equilibria of the game. We point out that when $L = 0$, and there are no network and capacity constraints (and of course, the network is connected), the problem corresponds to the classic setting of first best auction theory (see Fig. 4.3). **What can we say about this game? Is there an equilibrium? Is it unique? Can the agents 'game' the system?**

4.4.4   The Optimal Mechanism Design Problem

In this section, we assume that every participant knows the demand. As in 4.4.3, only producer i knows his true type c_i. The other agent and the ISO only know f_i. In order to decrease the market power of the agents and increase social welfare, we reverse the role of the principal (the ISO) and the agents, i.e., the principal 'bids' a contract to the agent. The contract should associate each bid profile $(b_i)_i$ with two vectors q and x, where q_i is the quantity of electricity agent i has to produce and x_i is amount of money he will receive. This contract is communicated to the producers before the bidding phase. Of course, this contract has to be incentive compatible; i.e., the payments described by the principal need to be high enough to make the agent willing to stay in the market. In this situation, the problem we are solving is the design of the optimal contract:

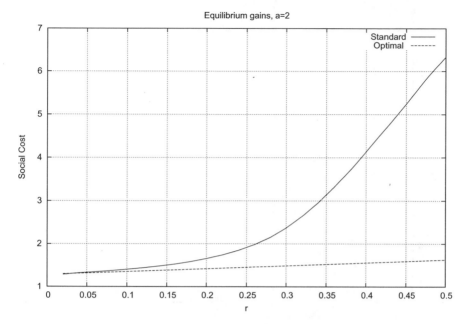

Fig. 4.3 Average total cost for the ISO (in the market described in Fig. 4.3) as a function of the loss coefficient r for the standard mechanism and the optimal mechanism. We take $f_1(c) = f_2(c) = 2c + (1 - \frac{2}{4})1_{c \leq \frac{1}{2}} + -2c + (1 + \frac{3}{2})1_{c \geq \frac{1}{2}}$. Note how r influences the social cost in the standard mechanism. The agents' market power increases with r. When r goes to zero, the two mechanisms lead to the same social cost. When $r = 0$, we recover a classic result on first and second best auctions

$$
\begin{aligned}
& \underset{q_j, h_{i,k}, x_j}{\text{minimize}} \quad \sum_j \mathbb{E} x_j(c) \\
& \text{subject to} \quad q_j(c) + \sum_i h_{i,j} - h_{j,i} - \sum_i \frac{L_{i,j}(h_{i,j}) + L_{j,i}(h_{j,i})}{2} \geq d_j \\
& \qquad\qquad \mathbb{E} x_j(c) - c_j(q_j(c)) \geq \mathbb{E} x_j(b) - c_j(q_j(b)) \\
& \qquad\qquad \mathbb{E} x_j(c) - c_j(q_j(c)) \geq 0 \\
& \qquad\qquad h_{i,j}, x_j \geq 0,
\end{aligned}
\tag{4.5}
$$

where $\mathbb{E}$ denotes the mean operator with respect to the f_i's, c denote the vector of production cost functions, and the constraints should be verified for all c. We refer to [4] for a justification of the formulation. We point out that this is an optimization problem over a functional set (so infinite dimensional) with an infinite number of constraints. The solution of this problem is an optimal mechanism; i.e., based on the information to the ISO, it provides the allocation and payment rules (q, x) that minimize the expected payments to the producers. We display some results in Figs. 4.3 and 4.4. **How do we build such problem? How can we solve it? How much better is the social surplus with an optimal design?**

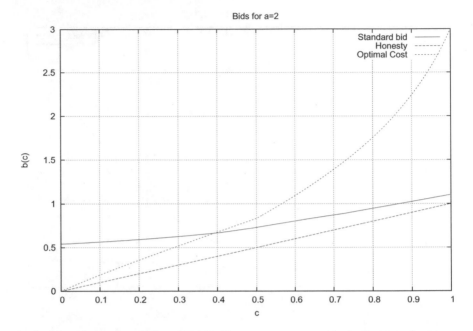

Fig. 4.4 Comparison of bidding strategy and information rent for the market described in Fig. 4.3. The standard bid strategy corresponds to the equilibrium strategy of the Bayesian game. The honesty strategy corresponds to a producer telling the truth. The optimal cost corresponds to the sum of the truth-telling strategy and the information rent

4.4.5 A Differential Equation

As noted in 4.4.3, the agents are playing a Bayesian Bertrand-like game in the standard setting. In this section, we propose a technique to compute a Nash equilibrium of this game. It is based on a fictitious play like dynamics. For instance, consider the simplified binodal, symmetric setting:

- Two agents;
- $L_{i,i'}(h_{i,i'}) = rh_{i,i'}$;
- $H = \mathbb{R}^2_+$;
- $\bar{q}_i = +\infty$;
- The cost functions and the bid functions are linear;
- $d_1 = d_2$: The demand is equal at each node.
- $f_1 = f_2 = f$.

We look for a symmetric equilibrium. If the agents iteratively change their bid functions proportionally to the corresponding increase in profit this will produce, the bid functions dynamics should be described by this formal differential equation.

$$\partial_t b(c, t) = \partial_b \pi_b(c, b(c, t)) \tag{4.6}$$

with

$$\pi_b(c, s) = \int_{C_{-i}} (s - c)(q_i(s, b(c_{-i}))f(c_{-i})dc_{-i}. \tag{4.7}$$

Is this dynamical system well posed? What conclusions can we draw from its study? Can we build such dynamics for more general settings?

4.5 Important Results

In this section, we sum up some results concerning the setting introduced previously. Most of the results focus on quadratic externalities (i.e., $L_{i',i}(h_{i',i}) = rh_{i',i}^2$) which is a realistic and simple assumption. Escobar and Jofré demonstrate in [1] the existence of noncooperative and Walrasian equilibrium when the ISO solves the standard ISO problem and demand is uncertain. The paper finishes with a welfare theorem for wholesale electricity auction. Escobar and Jofré give in [2] a lower bound on the market power exercised by each producer. The existence of a mixed strategy Nash equilibrium is given. The authors also give some regularity property on the ISO response function (condition to be a singleton, continuity, and Lipschitzianity). The cost functions are general. Figueroa, Jofré and Heymann study in [4], a binodal symmetric market with linear production cost functions and quadratic losses (i.e. $L_{i,i'}(h_{i,i'}) = rh_{i,i'}^2$, see Fig. 4.3). The principal minimal cost production plan problem was already solved in [2] and an explicit solution given. If we define

$$F(x, y) = d + \frac{1}{2r}\left(\frac{x-y}{x+y}\right)^2 - \frac{1}{r}\left(\frac{x-y}{x+y}\right) \quad \text{and} \quad \overline{q} = 2\left[\frac{1 - \sqrt{1 - 2dr}}{r}\right], \tag{4.8}$$

then the solution of the standard allocation problem is

$$q_i(c_i, c_{-i}) = \begin{cases} F(c_i, c_{-i}) & \text{if } F(c_i, c_{-i}) \geq 0 \text{ and } F(c_{-i}, c_i) \geq 0 \\ \overline{q} & \text{if } F(c_{-i}, c_i) < 0 \text{ and } F(c_i, c_{-i}) \geq 0 \\ 0 & \text{if } F(c_i, c_{-i}) < 0 \text{ and } F(c_{-i}, c_i) \geq 0 \end{cases} \tag{4.9}$$

This solution is used to compute an explicit solution of the mechanism design problem. The mechanism design solution is then compared to the standard setting for which numerical simulations are performed. The authors assume that the function $J_i : c_i \longrightarrow c_i + \frac{F_i(c_i)}{f_i(c_i)}$ is increasing in c_i, where f_i is the distribution of the marginal cost of producer i and F_i is its integral. Then, the main result is

Proposition 4.1 *Define*

$$\tilde{q} = 2\left[\frac{1 - \sqrt{1 - 2dr}}{r}\right]$$

Then, an optimal mechanism is given by

$$\hat{q}_i(c) = \begin{cases} F(J_i(c_i), J_{-i}(c_{-i})) & \text{if } F(J_i(c_i), J_{-i}(c_{-i})) \geq 0 \\ & \text{and } F(J_{-i}(c_{-i}), J_i(c_i)) \geq 0 \\ 0 & \text{if } F(J_i(c_i), J_{-i}(c_{-i})) \leq 0 \\ \tilde{q} & \text{if } F(J_{-i}(c_{-i}), J_i(c_i)) \leq 0 \end{cases}$$

$$\hat{h}_i(c) = \begin{cases} \frac{1}{r}\left[\frac{J_{-i}(c_{-i}) - J_i(c_i)}{J_{-i}(c_{-i}) + J_i(c_i)}\right] & \text{if } J_i(c_i) \leq J_{-i}(c_{-i}) \text{ and } F(J_{-i}(c_{-i}), J_i(c_i)) \geq 0 \\ \tilde{q} - 1 & \text{if } J_i(c_i) \leq J_{-i}(c_{-i}) \text{ and } F(J_{-i}(c_{-i}), J_i(c_i)) \leq 0 \\ 0 & \text{if not} \end{cases}$$

$$\hat{x}_i(c) = c_i \hat{q}_i(c) + \int_{c_i}^{\bar{c}_i} q_i(s, c_{-i}) ds$$

We point out that the mechanism is built with the standard ISO response function, where we just replace c_i by $J_i(c_i)$. The mechanism is incentive compatible; i.e., for any agent of any type, bidding the true type ensures a better profit for the agent than any other bidding strategy. Also the mechanism should satisfy a participation constraint, so that any agent can make a nonnegative profit. The optimal mechanism minimizes the total expected payment from the ISO to the agents while satisfying the incentive compatibility constraints, the participation constraints, and the nodal constraints.

4.6 The ISO Response for a Binodal Setting with Piecewise Linear Cost

4.6.1 Introduction

In this section, we derive an explicit expression for a specific example of ISO allocation problem as defined in 4.4.2. We study the binodal market with quadratic externalities displayed in Fig. 4.3. The production cost functions of both agents are made of two linear pieces, with a slope change when the production level is equal to $\bar{q}$. We denote by c_1 (resp. c_2) producer 1 (resp. 2) marginal cost when his production level is below $\bar{q}$ and by $\bar{c}_1$ (resp. $\bar{c}_2$) when it is above. The production cost functions are convex; i.e., $c_i < \bar{c}_i$, and the demand d is the same at both nodes. We end up with the following formulation for the ISO allocation problem:

$$\begin{aligned}
\underset{q_i, \bar{q}_i, h}{\text{minimize}} \quad & c_1 q_1 + \bar{c}_1 \bar{q}_1 + c_2 q_2 + \bar{c}_2 \bar{q}_2 \\
\text{subject to} \quad & q_i + \bar{q}_i + (-1)^i h \geq \frac{r}{2}(h^2) + d \quad (\lambda_i) \text{ for } i = 1, 2 \\
& q_i, \bar{q}_i \geq 0 \quad\quad\quad\quad\quad\quad\quad (\mu_i) \text{ for } i = 1, 2 \\
& q_i \leq \bar{q} \quad\quad\quad\quad\quad\quad\quad\quad (\gamma_i) \text{ for } i = 1, 2.
\end{aligned} \qquad (4.10)$$

In this formulation, q_i is the quantity produced by agent i at marginal cost c_i, and $\bar{q}_i$ is the quantity produced by i at marginal cost $\bar{c}_i$. These quantities are subject to positivity constraints with multipliers μ_i and $\bar{\mu}_i$. We also introduce λ_i the multipliers of the nodal constraints, and γ_i the multipliers of the constraints $q_i \leq \bar{q}$. We denote

$$F(\lambda_1, \lambda_2) = d + \frac{1}{r}\frac{\lambda_2 - \lambda_1}{\lambda_1 + \lambda_2} + \frac{1}{2r}\left(\frac{\lambda_2 - \lambda_1}{\lambda_1 + \lambda_2}\right)^2, \tag{4.11}$$

$$P(h) = h + \frac{rh^2}{2} + d, \tag{4.12}$$

$$k(\lambda_1, \lambda_2) = P\left(\frac{\lambda_2 - \lambda_1}{r(\lambda_1 + \lambda_2)}\right), \tag{4.13}$$

and

$$q_i^{tot} = q_i + \bar{q}_i. \tag{4.14}$$

We assume without loss of generality that $\bar{q} < 2d$ and $c_1 < c_2$. It is clear that if $q_1 < \bar{q}$, then $\bar{q}_1 = 0$. To solve this problem, we check whether $d < \bar{q}$ or $d \geq \bar{q}$.

4.6.2   If $d < \bar{q}$

By hypothesis $c_1 < c_2$. This implies that $q_1 \geq q_2$. So $\bar{q}_2 > 0$ implies that $q_1 = q_2 = \bar{q} > d$, which is not optimal. Therefore, we can set $\bar{q}_2 = 0$. We can also relax the constraint $q_2 < \bar{q}$ because it will not be binding for the optimal solution. So we rewrite the problem

$$\underset{q_i, \bar{q}_1, h}{\text{minimize}} \quad c_1 q_1 + \bar{c}_1 \bar{q}_1 + c_2 q_2$$

subject to

$$q_1 + \bar{q}_1 - h \geq \frac{r}{2}(h^2) + d \qquad (\lambda_1)$$

$$q_2 + h \geq \frac{r}{2}(h^2) + d \qquad (\lambda_2)$$

$$q_1, q_2, \bar{q}_1 \geq 0 \qquad (\mu_i)$$

$$q_1 \leq \bar{q} \qquad (\gamma_1)$$

The first-order conditions give

$$c_1 - \lambda_1 - \mu_1 + \gamma_1 = 0 \tag{4.15}$$

$$c_2 - \lambda_2 - \mu_2 = 0 \tag{4.16}$$

$$\bar{c}_1 - \lambda_1 - \bar{\mu}_1 = 0 \tag{4.17}$$

$$h = \frac{\lambda_2 - \lambda_1}{r(\lambda_1 + \lambda_2)} \tag{4.18}$$

There are four possible cases.

4.6.2.1   Case 1: $P(\frac{c_2-c_1}{r(c_1+c_2)}) \le \bar{q}$

We consider a relaxation of the problem by removing the constraint $q_1 \le \bar{q}$. In this relaxed problem, any optimal solution should verify $\bar{q}_1 = 0$ so the relaxed problem is equivalent to the linear cost functions allocation problem with costs c_i, for which we have an explicit formula of the solution. **We then notice that the optimal solution of the relaxed problem is admissible, so it is also the solution of** (4.10).

4.6.2.2   Case 2: $P(\frac{c_2-c_1}{r(c_1+c_2)}) > \bar{q}$ and $P(\frac{c_2-\bar{c}_1}{r(\bar{c}_1+c_2)}) \le \bar{q}$

We show that $\bar{q}_1 = 0$ and $q_1 = \bar{q}$.
If $\bar{q}_1 > 0$, then by complementarity of the multiplier $\bar{\mu}_1 = 0$, so with (4.17) $\lambda_1 = \bar{c}_1$. So by (4.18), we have $h = \frac{\lambda_2 - \bar{c}_1}{r(\bar{c}_1 + \lambda_2)}$. Then, by hypothesis and the fact that P is increasing and $\lambda_2 \le c_2$ we have that $P(h) \le P(\frac{c_2-\bar{c}_1}{r(\bar{c}_1+c_2)}) \le \bar{q}$. So $q_1 + \bar{q}_1 = q^{tot} \le \bar{q}$. Then using the fact that $q_1 < \bar{q} \Rightarrow \bar{q}_1 = 0$, we deduce that $\bar{q}_1$ is null, which is not the hypothesis. We conclude that $\bar{q}_1 = 0$.
By hypothesis, $\bar{q} < 2d$ and less than $\bar{q}$ is produced at node 1. Summing the two nodal constraints, we see that $q_2 > 0$, so that $\lambda_2 = c_2$.
Now if $q_1 < \bar{q}$, then by complementarity of the multiplier $\gamma_1 = 0$, and then with (4.15) $c_1 = \lambda_1$ and with (4.18), $h = \frac{c_2-c_1}{r(c_1+c_2)}$. Therefore, we get $q_1 + \bar{q}_1 = q_1^{tot} \ge P(\frac{c_2-c_1}{r(c_1+c_2)})$ (by the first nodal constraint), so by hypothesis $q_1^{tot} \ge \bar{q}$ and so $\bar{q}_1$ which we know as false. So $q_1 = \bar{q}$.

4.6.2.3   Case 3: $P(\frac{c_2-c_1}{r(c_1+c_2)}) > \bar{q}$ and $P(\frac{c_2-\bar{c}_1}{r(\bar{c}_1+c_2)}) > \bar{q}$ and $P(\frac{\bar{c}_1-c_2}{r(\bar{c}_1+c_2)}) > 0$

We show that $q_2 > 0$, $q_1 = \bar{q}$, $\bar{q}_1 > 0$ and $q_2 = P(\frac{\bar{c}_1-c_2}{r(\bar{c}_1+c_2)})$.
First we show that $q_2 > 0$. If $q_2 = 0$, then by the second nodal constraint $h \ge \frac{rh^2}{2} + d$, which means that $P(-h) \le 0$. Moreover, $\bar{q}_1 > 0$ because $2d > \bar{q}$. So by (4.17) $\lambda_1 = \bar{c}_1$. With (4.16) and (4.18), we have $P(\frac{\bar{c}_1-c_2}{r(\bar{c}_1+c_2)}) \le P(\frac{\bar{c}_1-\lambda_2}{r(\bar{c}_1+\lambda_2)}) = P(-h) \le 0$, which is false by hypothesis. So $q_2 > 0$. We deduce from this and (4.16) that $\lambda_2 = c_2$.

If $q_1 < \bar{q}$, then by complementarity of the multiplier $\gamma_1 = 0$, so by (4.15), $\lambda_1 = c_1$ and by (4.18), $h = \frac{c_2 - c_1}{r(c_1 + c_2)}$. Then, we get $q_1^{tot} \geq P(\frac{c_2 - c_1}{r(c_1 + c_2)})$ (by the first nodal constraint), so by hypothesis $q_1^{tot} \geq \bar{q}$, which implies $q_1 = \bar{q}$, which is absurd since we assumed $q_1 < \bar{q}$. So $q_1 = \bar{q}$.

If $\bar{q}_1 = 0$ then with (4.17), $\lambda_1 \leq \bar{c}_1$ and so with (4.18), $h \geq \frac{c_2 - \bar{c}_1}{r(\bar{c}_1 + c_2)}$. We then deduce by nodal constraint 1 and the hypothesis that $q_1^{tot} \geq P(h) \geq \bar{q}$, which implies that $\bar{q}_1 > 0$, which is absurd. So $\bar{q} > 0$.

We know that $\bar{q}_2 = 0$. Using the second nodal constraint, we get $q_2 = P(\frac{\bar{c}_1 - c_2}{r(\bar{c}_1 + c_2)})$. With the first nodal constraint, we have $\bar{q}_1 = P(\frac{c_2 - \bar{c}_1}{r(\bar{c}_1 + c_2)}) - \bar{q}$.

4.6.2.4   Case 4: $P(\frac{c_2 - c_1}{r(c_1 + c_2)}) > \bar{q}$ and $P(\frac{c_2 - \bar{c}_1}{r(\bar{c}_1 + c_2)}) > \bar{q}$ and $P(\frac{\bar{c}_1 - c_2}{r(\bar{c}_1 + c_2)}) \leq 0$

We show that $q_2 = 0$.
Indeed, if $q_2 > 0$, then $\lambda_2 = c_2$. Using the same reasoning as the one used in the third case, we would show that $q_1 = \bar{q}$ and $\bar{q}_1 > 0$. So that $h = -\frac{\bar{c}_1 - c_2}{r(\bar{c}_1 + c_2)}$. So using $P(\frac{\bar{c}_1 - c_2}{r(\bar{c}_1 + c_2)}) \leq 0$, we see that nodal constraint 2 is satisfied with $q_2 = 0$, so the solution is not optimal, which is absurd. So $q_2 = 0$.

4.6.2.5   We Conclude

Theorem 4.1 *Assuming $d < \bar{q} < 2d$, then:*

$$q_1^{tot} = k(c_1, c_2) \text{ and } q_2^{tot} = k(c_2, c_1) \text{ if } k(c_1, c_2) \leq \bar{q}$$

$$q_1^{tot} = \bar{q} \text{ and } q_2^{tot} = \bar{q} - 2\frac{-1 + \sqrt{1 + 2r(\bar{q} - d)}}{r} \text{ if } k(c_1, c_2) > \bar{q} \text{ and } k(c_1, c_2) \leq \bar{q}$$

$$q_1^{tot} = k(\bar{c}_1, c_2) \text{ and } q_2^{tot} = k(c_2, \bar{c}_1) \text{ if } k(c_1, c_2) > \bar{q}, k(\bar{c}_1, c_2) > \bar{q} \text{ and } k(c_2, \bar{c}_1) > 0$$

$$q_1^{tot} = 2\frac{1 - \sqrt{1 - 2dr}}{r} \text{ and } q_2^{tot} = 0 \text{ if } k(c_1, c_2) > \bar{q}, k(\bar{c}_1, c_2) > \bar{q} \text{ and } k(c_2, \bar{c}_1) \leq 0$$

4.6.3   Case $d \geq \bar{q}$

Since we consider that c_1 is smaller than c_2, there are two possibilities. Either the $\bar{c}_i$ are all bigger than the c_i or $\bar{c}_1$ is smaller than c_2.

4.6.3.1   If the $\bar{c}_i$ Are All Bigger Than the c_i

In this case, we first show that $q_1 = q_2 = \bar{q}$. The problem then writes

$$\underset{q_i,\bar{q}_i,h}{\text{minimize}} \quad \bar{c}_1\bar{q}_1 + \bar{c}_2\bar{q}_2$$

$$\text{subject to} \quad \bar{q}_1 + -h \geq \frac{r}{2}(h^2) + d - \bar{q} \qquad (\lambda_1)$$

$$\bar{q}_2 + h \geq \frac{r}{2}(h^2) + d - \bar{q} \qquad (\lambda_2)$$

$$\bar{q}_i \geq 0 \qquad (\mu_i) \text{ for i} = 1, 2$$

which corresponds to the linear case problem with a demand of $d - \bar{q}$ and costs of $\bar{c}_i$.

4.6.3.2   If $\bar{c}_1$ is Smaller than c_2

We point out that replacing c_1 by $\bar{c}_1$ does not change the solution.

If $F(c_2, \bar{c}_1) \leq \bar{q}$, we show that $\bar{q}_2 = 0$ the problem can be reduced to the linear production cost problem with demand d and marginal costs $\bar{c}_1$ and c_2.

If $F(c_2, \bar{c}_1) > \bar{q}$, we show that we can reduce the linear production cost problem with demand $d - q$ and marginal costs $\bar{c}_1$ and $\bar{c}_2$.

Theorem 4.2 *If $d \geq \bar{q}$, then*

- *If $c_i \leq \bar{c}_j$ for all i, j, then we get the result by solving the linear problem with demand $d - \bar{q}$ and costs $\bar{c}_i$ and adding $\bar{q}$ to the quantity we get.*
- *If $0 \leq F(c_2, \bar{c}_1) \leq \bar{q}$, we reduce to the linear allocation problem with demand d and marginal cost $\bar{c}_1$ and c_2.*
- *If $F(c_2, \bar{c}_1) > \bar{q}$, we reduce the problem to the linear allocation problem with demand $d - q$ and marginal costs $\bar{c}_1$ and $\bar{c}_2$ and add $\bar{q}$ to the q_is we get.*

4.7   Ongoing Work

We are currently working on several questions raised in this chapter. In particular, we have shown that for a market with n-pieces piecewise linear production cost functions and any number of producers, there is a mechanism design with an explicit formulation.

4.8   Exercise

Consider the setting of Sect. 4.6, but with a linear production cost (instead of piece-wise linear).

(1) Assume agent 1 bids b_1 and agent 2 bids b_2. Compute the optimal allocation for the ISO as a function of r, d, and b_i.

(2) Compute the best reply of a producer of production cost c against a producer who bids b.

(3) Assume the production cost is c for both agents (and this is common knowledge). Compute the equilibrium strategies of the Bertrand game.

(4) What can we say about the market power of the producers in this setting?

Acknowledgements   The authors would like to thank Idalia Gonzalez for her English editing.

References

1. Escobar, J.F., Jofré, A.: Equilibrium Analysis of Electricity Auctions. Department of Economics Stanford University (2014)
2. Escobar, J.F., Jofré, A.: Monopolistic competition in electricity networks with resistance losses. Econ. Theor. **44**(1), 101–121 (2010)
3. Escobar, J.F., Jofré, A.: Equilibrium analysis for a network market model. In: Robust Optimization-Directed Design, pp. 63–72. Springer, Berlin (2006)
4. Figueroa, N., Jofré, A., Heymann, B.: Cost-Minimizing Regulations for a Wholesale Electricity Market (2015)
5. Klemperer, P.D., Meyer, M.A.: Supply function equilibria in oligopoly under uncertainty. Econometrica J. Econometric Soc. 1243–1277 (1989)
6. Anderson, E.J., Philpott, A.B.: Optimal offer construction in electricity markets. Math. Oper. Res. **27**(1), 82–100 (2002)
7. Anderson, E.J., Hu, X.: Finding supply function equilibria with asymmetric firms. Oper. Res. **56**(3), 697–711 (2008)
8. Anderson, E.J.: On the existence of supply function equilibria. Math. Program. **140**(2), 323–349 (2013)
9. Myerson, R.B.: Optimal auction design. Math. Oper. Res. **6**(1), 58–73 (1981)
10. Laffont, J.-J., Martimort, D.: The Theory of Incentives: The Principal-Agent Model. Princeton University Press, USA (2009)
11. Aussel, D., Correa, R., Marechal, M.: Electricity spot market with transmission losses. Management **9**(2), 275–290 (2013)
12. Hu, X., Ralph, D.: Using epecs to model bilevel games in restructured electricity markets with locational prices. Oper. Res. **55**(5), 809–827 (2007)
13. Gabriel, S.A., Conejo, A.J., Fuller, J.D., Hobbs, B.F., Ruiz, C.: Complementarity modeling in energy markets. International series in Operations Research & Management Science (2013)

Chapter 5
Reflection Methods for Inverse Problems with Applications to Protein Conformation Determination

Jonathan M. Borwein and Matthew K. Tam

Abstract The Douglas–Rachford reflection method is a general-purpose algorithm useful for solving the feasibility problem of finding a point in the intersection of finitely many sets. In this chapter, we demonstrate that applied to a specific problem, the method can benefit from heuristics specific to said problem which exploit its special structure. In particular, we focus on the problem of protein conformation determination formulated within the framework of matrix completion, as was considered in a recent paper of the present authors.

Keywords Reflection methods · Inverse problems · Protein conformation

5.1 Techniques of Variational Analysis

This chapter builds on a series of seven lectures titled *Techniques of Variational Analysis* given by the first author at the CIMPA school *Generalized Nash Equilibrium Problems, Bilevel Programming and MPEC* held on November 25 to December 6, 2013, University of Delhi, New Delhi, India. In this written presentation, we focus on *reflection methods* for *protein conformation determination*, as was discussed in the seventh and final lecture of the series. The complete lectures—one through six taken from [14]—can be found online at:

http://www.carma.newcastle.edu.au/jon/ToVA/links.html

Before turning our attention to reflection methods, we briefly outline the content of the first six lectures.

J. M. Borwein · M. K. Tam (✉)
Institut für Numerische und Angewandte Mathematik, Universität Göttingen,
37085 Göttingen, Germany
e-mail: m.tam@math.unigoettingen.de

J. M. Borwein
e-mail: jon.borwein@gmail.com

© Springer Nature Singapore Pte Ltd. 2017
D. Aussel and C. S. Lalitha (eds.), *Generalized Nash Equilibrium
Problems, Bilevel Programming and MPEC*, Forum for Interdisciplinary
Mathematics, https://doi.org/10.1007/978-981-10-4774-9_5

- **Lectures 1 & 2** provided an introduction to variational analysis and variational principles [14, §1–§2].
- **Lectures 3 & 4** introduced nonsmooth analysis: normal cones and subdifferentials of lower semi-continuous functions, Fréchet and limiting calculus [14, §3.1–§3.4], and discussed convex functions and their calculus rules [14, §4.1–§4.4].
- **Lecture 5** turned to multifunction analysis: sequences of sets, continuity of maps, minimality and maximal monotonicity, and distance functions [14, §5.1–§5.3].
- **Lecture 6** focussed on convex feasibility problems and the method of alternating projections [14, §4.7], therefore providing the preliminary background for the rest of this chapter.

5.2   Introduction to Reflection Methods

Given a (finite) family of sets, the corresponding *feasibility problem* is to find a point contained in their intersection. *Douglas–Rachford reflection methods* form a class of general-purpose iterative algorithms which are useful for solving such problems. At each iteration, these methods perform *(metric) reflections* and *(metric/nearest point) projections* with respect to the individual constraint sets in a prescribed fashion. Such methods are most useful when applied to feasibility problems whose constraint sets have more easily computable reflections and projections than does the intersection.

When the underlying constraint sets are all convex, Douglas–Rachford methods are relatively well understood [6, 7, 12, 13]—their behavior can be analyzed using nonexpansivity properties of convex projections and reflections. In the absence of convexity, recent result has assumed the constraint sets to possess other structural and regularity properties [1, 11, 21]. However, at present, this developing theoretical foundation is not sufficiently rich to explain many of the successful applications in which one or more of the constraint sets lack convexity [2, 3, 18, 19]. In these cases, the method can be viewed as a heuristic inspired by its behavior within fully convex settings.

More generally, with any algorithm there is typically a trade-off between the scope of their applicability and tailoring of performance to particular instances. Douglas–Rachford reflection methods are no different. Owing to these methods' broad applicability, potential for further problem-specific refinements when applied to special classes of feasibility problems is possible.

In this chapter, we investigate and develop one such refinement with a focus on application of the Douglas–Rachford method to *protein conformation determination*. This application was previously considered as part of [3]. We now propose problem-specific heuristics and also study the effect of increasing problem size. We finish by demonstrating a complementary application of the approach arising in the context of *ionic liquid chemistry*.

The remainder of this chapter is organized as follows. In Sects. 5.3, 5.4, 5.5, and 5.6, we introduce the necessary mathematical preliminaries along with the Douglas–Rachford reflection method, before formulating the protein conformation determination problem. Substantial, numerical, and graphical results are given in Sect. 5.7 and concluding remarks in Sect. 5.8.

5.3 Mathematical Preliminaries

Let $\mathbb{E}$ denote a Euclidean space, that is, a finite dimensional Hilbert space. We will mainly be concerned with the space $\mathbb{R}^{m \times m}$ (i.e., real $m \times m$ matrices) equipped with the inner product given by

$$\langle A, B \rangle := \operatorname{tr}(A^T B).$$

Here, the symbol $\operatorname{tr}(X)$ (resp. X^T) denotes the trace (resp. transpose) of the matrix X. The induced norm is the *Frobenius norm* and can be expressed as

$$\|A\|_F := \sqrt{\operatorname{tr}(A^T A)} = \sqrt{\sum_{i=1}^{m} \sum_{j=1}^{m} a_{ij}^2}.$$

The subspace of real symmetric $m \times m$ matrices is denoted S^m and the cone of positive semi-definite $m \times m$ matrices by S_+^m.

Given sets $C_1, C_2, \ldots, C_N \subseteq \mathbb{E}$, the *feasibility problem* is

$$\text{find } x \in \bigcap_{i=1}^{N} C_i. \tag{5.1}$$

When the intersection in (5.1) is empty, one often seeks a "good" surrogate for a point in the intersection. When $N = 2$, a useful surrogate is a pair of points, one from each set, which minimize the distance between the sets—a *best approximation pair* [6].

5.4 Matrix Completion

A *partial (real) matrix* is an $m \times m$ array for which entries only in certain locations are known. Given a partial matrix $A = (a_{ij}) \in \mathbb{R}^{m \times m}$, a matrix $B = (b_{ij}) \in \mathbb{R}^{m \times m}$ is a *completion* of A if $b_{ij} = a_{ij}$ whenever a_{ij} is known. The problem of *(real) matrix completion* is the following: *Given a partial matrix, find a completion belonging to a specified family of matrices.*

Matrix completion can be naturally formulated as a feasibility problem. Let A be the partial matrix to be completed. Choose $C_1, C_2, \ldots, C_N$ such that their intersection is equal to the intersection of completions of A with the specified matrix family. Then, (5.1) is precisely the problem of matrix completion for A. The simplest such case is when C_1 is the set of all completions of A and the intersection of $C_2, \ldots, C_N$ equals the desired matrix class.

Remark 5.1 More generally, one may profitably consider matrix completion for rectangular matrices [3], for example, with doubly stochastic matrices. However, since the partial matrices in the discussed protein application are always square, for the purposes of this discussion, we only concern ourselves with the square case.

5.5   The Douglas–Rachford Reflection Method

The *projection onto* $C \subseteq \mathbb{E}$ is the set-valued mapping $P_C : \mathbb{E} \rightrightarrows C$ which maps any point $x \in \mathbb{E}$ to its sets of nearest points in C. More precisely,

$$P_C(x) = \left\{ c \in C : \|x - c\| \leq \inf_{y \in C} \|x - y\| \right\}.$$

The *reflection with respect to* C is the set-valued mapping $R_C : \mathbb{E} \rightrightarrows \mathbb{E}$ given by $R_C = 2P_C - I$, where I denotes the identity mapping. An illustration is shown in Fig. 5.1.

When C is nonempty, closed, and convex, its corresponding projection operator (and hence its reflection) is single-valued (see, e.g., [16, Chap. 1.2]).

Given $A, B \subseteq \mathbb{E}$ and $x_0 \in \mathbb{E}$, the *Douglas–Rachford reflection method* is the fixed-point iteration given by

$$x_{n+1} \in T_{A,B} x_n \text{ where } T_{A,B} = \frac{I + R_B R_A}{2}. \tag{5.2}$$

Fig. 5.1 (Left) The (single-valued) projection, p_i, and reflection, r_i, of the point x_i onto a convex set, for $i = 1, 2$. (Right) The (set-valued) projection, $\{p_1, p_2\}$, and reflection, $\{r_1, r_2\}$, of the point x onto a nonconvex set. Note the nonexpansivity of the reflection in the convex case

We refer to the sequence $(x_n)_{n=1}^{\infty}$ as a *Douglas–Rachford sequence* and to the mapping $T_{A,B}$ as the *Douglas–Rachford operator*.

We now recall the behavior of the Douglas–Rachford method in the classical convex setting. In this case, $T_{A,B}$ is single-valued as a consequence of the single-valuedness of each of P_A, P_B, R_A and R_B. We denote the set of *fixed points* of a single-valued mapping T by Fix $T = \{x \in \mathbb{E} : Tx = x\}$ and the *normal cone* of a convex set C at the point x by

$$N_C(x) = \begin{cases} \{y \in \mathbb{E} : \langle C - x, u \rangle \leq 0\} & \text{if } x \in C, \\ \emptyset & \text{otherwise.} \end{cases}$$

For convenience, we also introduce the two sets

$$E = \left\{ x \in A : \inf_{a \in A} \|a - x\| \leq \inf_{a \in A, b \in B} \|a - b\| \right\},$$

$$F = \left\{ x \in B : \inf_{b \in B} \|x - b\| \leq \inf_{a \in A, b \in B} \|a - b\| \right\},$$

and the vector $v = P_{\overline{B-A}}(0)$. Here, the overline denotes the closure of the set.

Theorem 5.1 (Convex Douglas–Rachford in finite dimensions [6]). *Suppose $A, B \subseteq \mathbb{E}$ are closed and convex. For any $x_0 \in \mathbb{E}$, define $x_{n+1} = T_{A,B}x_n$. Then, there is some $v \in \mathbb{E}$ such that:*

(i) *$x_{n+1} - x_n = P_B R_A x_n - P_A x_n \to v$ and $P_B P_A x_n - P_A x_n \to v$.*
(ii) *If $A \cap B \neq \emptyset$, then $(x_n)_{n=1}^{\infty}$ converges to a point in*

$$\text{Fix}(T_{A,B}) = (A \cap B) + N_{\overline{A-B}}(0);$$

otherwise, $\|x_n\| \to +\infty$.
(iii) *Exactly one of the following two alternatives holds.*

(a) *$E = \emptyset$, $\|P_A x_n\| \to +\infty$, and $\|P_B P_A x_n\| \to +\infty$.*
(b) *$E \neq \emptyset$, the sequences $(P_A x_n)_{n=1}^{\infty}$ and $(P_B P_A x_n)_{n=1}^{\infty}$ are bounded, and their cluster points belong to E and F, respectively; in fact, the cluster points of*

$$((P_A x_n, P_B R_A x_n))_{n=1}^{\infty} \text{ and } ((P_A x_n, P_B P_A x_n))_{n=1}^{\infty}$$

are a best approximation pairs relative to (A, B).

Theorem 5.1 provides the template for application of the Douglas–Rachford method as a heuristic for nonconvex feasibility problems. Furthermore, this theorem also shows that for the Douglas–Rachford method the sequence of primary interest is not the fixed-point iterates $(x_n)_{n=1}^{\infty}$ themselves, but their *shadows* $(P_A x_n)_{n=1}^{\infty}$ (Fig. 5.2).

Remark 5.2 (Douglas–Rachford splitting). The Douglas–Rachford reflection method can be viewed as a special case of the *Douglas–Rachford splitting algorithm* for finding a zero of the sum of two maximally monotone operators. This more general splitting method iterates by using *resolvents* of the given maximally monotone operators rather than projection operators of sets. The reflection method is obtained in the special case in which the maximal monotone operators are normal cones to the feasibility problem sets. For details, we refer the reader to [5].

Within an implementation of the Douglas–Rachford method, computation of the projection operators are the component typically requiring the most resources. It is therefore beneficial to store two additional sequences in memory: the *shadow sequence* $(P_A x_n)_{n=1}^{\infty}$ and the sequence $(P_B R_A x_n)_{n=1}^{\infty}$. This is because iteration (5.2) is expressible as

$$x_{n+1} \in x_n + P_B R_A x_n - P_A x_n$$
$$= x_n + P_B(2P_A x_n - x_n) - P_A x_n. \tag{5.3}$$

An implementation utilizing this approach is given in Fig. 5.3. The stopping criterion uses a relative error and is discussed in Sect. 5.7.

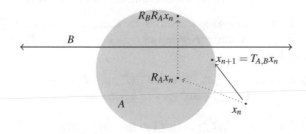

Fig. 5.2 One iteration of the Douglas–Rachford method for the sets $A = \{x \in \mathbb{E} : \|x\| \leq 1\}$ and $B = \{x \in \mathbb{E} : \langle a, x \rangle = b\}$

```
Input: x₀ ∈ 𝔼 and ε > 0
n = 0;
p₀ ∈ P_A(x₀);
while n = 0 or ‖rₙ − pₙ‖ > ε‖pₙ‖ do
    rₙ ∈ P_B(2pₙ − xₙ);
    xₙ₊₁ = xₙ + rₙ − pₙ;
    pₙ₊₁ ∈ P_A(xₙ₊₁);
    n = n + 1;
end
Output: pₙ ∈ 𝔼
```

Fig. 5.3 Implementation of the basic Douglas–Rachford algorithm

5.6   Protein Conformation Determination

Proteins are large biomolecules which are comprised of multiple amino acid residues,[1] each of which typically consists of between 10 and 25 atoms. Proteins participate in virtually every cellular process, and knowledge of their structural conformation gives insight into the mechanisms by which they perform.

One of many techniques that can be used to determine conformation is *nuclear magnetic resonance (NMR)*. Currently, NMR is only able to nondestructively resolve relatively short distances (i.e., those less than $\sim 6\,\text{Å}$). In the proteins we consider, this corresponds to less than 9% of all nonzero inter-atom distances.

We now formulate the problem of protein conformation determination as a computationally tractable matrix completion problem. In fact, our formulation is a *low-rank Euclidean distance matrix completion problem*. We next introduce the necessary definitions.

We say that a matrix $D = (D_{ij}) \in \mathbb{R}^{m \times m}$ is a *Euclidean distance matrix (EDM)* if there exists points $z_1, z_2, \ldots, z_n \in \mathbb{R}^m$ such that

$$D_{ij} = \|z_i - z_j\|^2 \text{ for } i, j = 1, 2, \ldots, m. \tag{5.4}$$

Clearly, any EDM is symmetric, nonnegative, and hollow (i.e., contains only zeros along its main diagonal). When (5.4) holds for a set of points in $\mathbb{R}^q$, we say D is *embeddable* in $\mathbb{R}^q$. If D is embeddable in $\mathbb{R}^q$ but not in $\mathbb{R}^{q-1}$, then we say that D is *irreducibly embeddable* in $\mathbb{R}^q$.

We now recall a useful characterization of EDMs, due to Hayden and Wells [20]. In what follows, the matrix $Q \in \mathbb{R}^{m \times m}$ is the *Householder matrix* given by

$$Q = I - \frac{2vv^T}{v^T v}, \text{ where } v = \begin{bmatrix} 1, & 1, & \ldots & 1, & 1 + \sqrt{m} \end{bmatrix}^T \in \mathbb{R}^m.$$

Theorem 5.2 (EDM characterization [20, Th. 3.3]). *A nonnegative, symmetric, hollow matrix* $X \in \mathbb{R}^{m \times m}$ *is a Euclidean distance matrix if and only if the block* $\widehat{X} \in \mathbb{R}^{(m-1) \times (m-1)}$ *in*

$$Q(-X)Q = \begin{bmatrix} \widehat{X} & d \\ d^T & \delta \end{bmatrix} \tag{5.5}$$

is positive semi-definite. In this case, X is irreducibly embeddable in $\mathbb{R}^q$ where $q = \text{rank}(\widehat{X}) \leq m - 1$.

The problem of *low-rank Euclidean distance matrix completion* can now be formulated. Let D denote a partial Euclidean distance matrix, with entry D_{ij} known whenever $(i, j) \in \Omega$ for some index set Ω, which is embeddable in $\mathbb{R}^q$. Without loss of generality, we make the following three simplifying assumptions on the partial matrix D and index set Ω.

---

[1]When two amino acids form a peptide bond, a water molecule is formed. An *amino acid residue* is what remains of each amino acid after this reaction.

1. (nonnegative) $D \geq 0$ (i.e., $D_{ij} \geq 0$ for all $i, j = 1, 2, \ldots, m$);
2. (hollow) $D_{ii} = 0$ and $(i, i) \in \Omega$ for $i = 1, 2, \ldots, m$;
3. (symmetric) $(i, j) \in \Omega \iff (j, i) \in \Omega$, and $D_{ij} = D_{ji}$ for all $(i, j) \in \Omega$.

We define two constraint sets

$$C_1 = \left\{ X \in S^m : X \geq 0, \ X_{ij} = D_{ij} \text{ for all } (i, j) \in \Omega \right\},$$

$$C_2 = \left\{ X \in S^m : Q(-X)Q = \begin{bmatrix} \widehat{X} & d \\ d^T & \delta \end{bmatrix}, \ \begin{matrix} \widehat{X} \in S_+^{m-1}, \ d \in \mathbb{R}^{m-1} \\ \text{rank} \widehat{X} \leq q, \ \delta \in \mathbb{R} \end{matrix} \right\}. \quad (5.6)$$

In light of Theorem 5.2, the problem of *low-rank Euclidean distance matrix completion* can be cast as the two-set feasibility problem

$$\text{find } X \in C_1 \cap C_2 \subseteq S^m.$$

That is, a matrix X is a low-rank Euclidean distance matrix which completes D if and only if $X \in C_1 \cap C_2$. Some comments regarding the constraint sets in (5.6) are in order.

The set C_1 encodes the experimental data obtained from NMR and the a priori knowledge that the matrix is nonnegative, symmetric, and hollow. Its projection has a simple formulae, as we now show.

Proposition 5.1 (Projection onto C_1). *Let $X \in \mathbb{R}^{m \times m}$. Then, $P_{C_1} X$ is given element-wise by*

$$(P_{C_1} X)_{ij} = \begin{cases} D_{ij}, & (i, j) \in \Omega \\ \max\{0, X_{ij}\}, & (i, j) \notin \Omega \end{cases} \quad \text{for } i, j = 1, 2, \ldots, m.$$

Proof Let Y be any matrix in C_1, and let P be the matrix given by the proposed projection formula (clearly $P \in C_1$). Then

$$\|X - Y\|_F^2 = \sum_{(i,j) \in \Omega} (X_{ij} - Y_{ij})^2 + \sum_{\substack{(i,j) \notin \Omega \\ \text{s.t. } X_{ij} < 0}} (X_{ij} - Y_{ij})^2 + \sum_{\substack{(i,j) \notin \Omega \\ \text{s.t. } X_{ij} \geq 0}} (X_{ij} - Y_{ij})^2$$

$$\geq \sum_{(i,j) \in \Omega} (X_{ij} - D_{ij})^2 + \sum_{\substack{(i,j) \notin \Omega \\ \text{s.t. } X_{ij} < 0}} (X_{ij} - 0)^2 + \sum_{\substack{(i,j) \notin \Omega \\ \text{s.t. } X_{ij} \geq 0}} (X_{ij} - X_{ij})^2 \quad (5.7)$$

$$= \|X - P\|_F^2.$$

Furthermore, observe that equality in (5.7) holds if and only if $Y = P$. Altogether, we have that $P \in C_1$ and

$$\|X - Y\|_F^2 > \|X - P\|_F^2 \text{ for all } Y \in C_1 \setminus \{P\},$$

which completes the proof. $\square$

Remark 5.3 Since C_1 is a closed convex set, an alternative (less direct) proof of Proposition 5.1 can be given using the standard variational characterization of convex projections [16, Th. 1.2.4].

Using the necessary condition given by Theorem 5.2, the nonconvex set C_2 encodes the a priori knowledge that the matrix of interest is a EDM together with the dimension of the space in which the corresponding points generating the matrix are contained. We now derive the projection onto C_2.

Theorem 5.3 (Nearest low-rank EDMs [3]). *Let $X \in S^m$ be a nonnegative, hollow matrix. Then*

$$P_{C_2}(X) = \left\{ -Q \begin{bmatrix} \widehat{Y} & d \\ d^T & \delta \end{bmatrix} Q : Q(-X)Q = \begin{bmatrix} \widehat{X} & d \\ d^T & \delta \end{bmatrix}, \begin{matrix} \widehat{X} \in S^{m-1}, \ d \in \mathbb{R}^{m-1} \\ \widehat{Y} \in P_M \widehat{X}, \ \delta \in \mathbb{R}; \end{matrix} \right\},$$

where M is the set of positive semi-definite matrices having rank q or less. In particular, $P_{C_2}(X)$ is a singleton if and only if $P_M \widehat{X}$ is a singleton.

Proof Let Y be any matrix in C_2. That is,

$$Q(-Y)Q = \begin{bmatrix} \widehat{Y} & c \\ c^T & \beta \end{bmatrix}, \quad \text{for some } c \in \mathbb{R}^{m-1}, \ \beta \in \mathbb{R}, \ \widehat{Y} \in M.$$

Using the orthogonality of Q, we compute

$$\begin{aligned}
\|X - Y\|_F^2 &= \|Q(X - Y)Q\|_F^2 = \|Q(-X)Q - Q(-Y)Q\|_F^2 \\
&= \left\| \begin{bmatrix} \widehat{X} & d \\ d^T & \delta \end{bmatrix} - \begin{bmatrix} \widehat{Y} & c \\ c^T & \beta \end{bmatrix} \right\|_F^2 = \left\| \begin{bmatrix} \widehat{X} - \widehat{Y} & (d - c) \\ (d - c)^T & (\delta - \beta) \end{bmatrix} \right\|_F^2 \quad (5.8) \\
&= \|\widehat{X} - \widehat{Y}\|_F^2 + 2\|d - c\|^2 + |\gamma - \beta|^2.
\end{aligned}$$

To complete the proof, we observe that (5.8) is minimized if and only if $c = d$, $\gamma = \beta$, and $\widehat{Y} \in P_M \widehat{X}$. $\qquad\square$

The set M in Theorem 5.3 is a set of low-rank positive semi-definite matrices. One method to compute its projection (and the one we will use) is by exploiting the eigen-decomposition of $\widehat{X}$. Denote by diag(λ) the diagonal matrix given by placing the elements of the vector $\lambda \in \mathbb{R}^m$ along the main diagonal. Let $\widehat{X} = U\text{diag}(\lambda)U^T$ be an eigen-decomposition (of $\widehat{X}$) with

$$\lambda_1 \geq \lambda_2 \geq \cdots \geq \lambda_q^+ \geq \cdots \geq \lambda_m.$$

A projection onto the set is then given by

$$U\text{diag}((\lambda_1^+, \lambda_2^+, \ldots, \lambda_q^+, 0 \ldots, 0, 0))U^T,$$

where x^+ denotes max$\{0, x\}$.

5.7   Computational Experiments

We apply the formulation of Sect. 5.6 to six proteins, shown in Table 5.1, obtained
from the *RCSB Protein Data Bank*[2] [8]. As part of [3], reconstructions of the same
six proteins were attempted using a partial EDM containing only distances less than
6 Å. Here, we attempt reconstructions using partial EDMs which, in addition to these
short-range distances, incorporate other a priori information. In particular, we include
inter-atomic distances greater than 6 Å for atoms from within the same residue in
the partial EDM. This is reasonable since the structure of the individual residues is
known. For 1PTQ, this information gives approximately a further 0.2% of the total
nonzero inter-atomic distances.

Our experiments were implemented in *Cython* and performed on a machine having an Intel Xeon E5540 @ 2.83 GHz running Red Hat Enterprise Linux 6.5. A
combination of the Cython platform and optimized code gave approximately a ten-
fold speed up compared to [3]. This allowed for a greater number of iterations to
be performed and hence the use of the more robust (albeit still heuristic) stopping
criterion given in Fig. 5.3 as opposed to simply performing a fixed number of iterations. The reconstructed EDM, x, was converted to points $z_1, z_2, \ldots, z_m \in \mathbb{R}^3$ using
the algorithm in Fig. 5.4.

Remark 5.4 It is worth emphasizing that our primary concern is the quality of the
reconstruction, rather than the time required to perform the reconstruction. This is
because, if done well, one only needs to determine the conformation once.

We report two error metrics, which we now explain. Denote the actual EDM by
x^{actual}. The first error metric is a measure of the *error in the reconstructed EDM* and
is given by

$$\text{EDM-error} = \|x^{\text{actual}} - x\|_F = \sqrt{\sum_{i,j=1}^{m} \left| x_{ij}^{\text{actual}} - x_{ij} \right|^2}.$$

Denote the actual atom positions by $z_1^{\text{actual}}, z_2^{\text{actual}}, \ldots, z_m^{\text{actual}} \in \mathbb{R}^3$. The second
error metric measures the *error in the reconstructed atom positions* $z_1, z_2, \ldots, z_m \in \mathbb{R}^3$. Since EDMs are invariant under translation, reflection, and rotation of the points
by which they are induced, we first perform a *Procrustes analysis* [17] to obtain
$\tilde{z}_1, \tilde{z}_2, \ldots, \tilde{z}_m \in \mathbb{R}^3$. These points are a best fit of the reconstructed points when the
aforementioned transformations are allowed. The second error metric is given by

$$\text{Position-error} = \sqrt{\sum_{k=1}^{m} \|z_k^{\text{actual}} - \tilde{z}_k\|_2^2}.$$

---

[2]RCSB Protein Data Bank: www.rcsb.org/pdb.

Table 5.1 Number of atoms, residues, known, and total nonzero inter-atomic distances in our six test proteins

Protein	Atoms	Residues	Total nonzero distances	Known nonzero distances (%)
1PTQ	404	50	81,406	8.9207
1HOE	581	74	168,490	6.4105
1LFB	641	99	205,120	5.6362
1PHT	988	85	236,328	4.6501
1POA	1067	118	568,711	3.6375
1AX8	1074	146	576,201	3.5606

```
Input: x ∈ X ;                          /* a Euclidean distance matrix */
L = I − ee^T/n where e = (1,1,...,1)^T;
τ = −LDL/2;
USV^T = SingularValueDecomposition(τ);
Z = first q columns of U√S;
z_i = ith row of Z for i = 1,2,...,m;
Output: z_1, z_2,..., z_q ∈ R^q ;        /* points corresponding to x */
```

Fig. 5.4 Conversion of EDM to points in $\mathbb{R}^q$

When comparing the relative size of these two errors, it is worth noting that the summation in the EDM-error contains m^2 terms, whereas the summation in the position-error contains only $3m$.

Remark 5.5 (Decibel error). It is also common to consider the relative error in *decibels (dB)*, as was reported in [3]. That is,

$$\text{Relative error (dB)} = 10 \log_{10} \left(\frac{\|P_B R_A x - P_A x\|_F^2}{\|P_A x\|_F^2} \right).$$

In this study, the relative error in decibels is not reported. This is unnecessary because the stopping criterion used in Fig. 5.3 is equivalent to requiring that the decibel error be less than $10 \log_{10}(\varepsilon^2)$. Requiring that $\varepsilon = 10^{-5}$ corresponds to aiming at a relative error of $-100\,\text{dB}$.

Remark 5.6 (Stopping criterion and tolerance). In the computational experiments that follow, the stopping tolerance is taken to be $\varepsilon = 10^{-5}$. We now provide some justification for this choice.

For each of the six proteins, Fig. 5.5 shows the relative error as a function of the number of iterations starting from a given initial point for the Douglas–Rachford method.

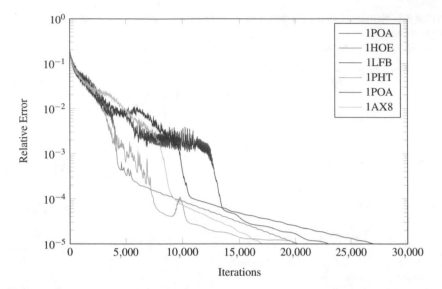

Fig. 5.5 The relative error as a function of iterations (vertical axis is logarithmic)

- When the number of iteration is less than 5000, the relative error exhibits non-monotone oscillatory behavior—which seems to provide much of the potency of the method. It seems to allow the reflection method to sample regions and avoid settling at an inferior local minimum of the configuration space. In [3], we observed that the alternating projection method, which is monotonic, fails to produce good reconstructions.
- When the relative error is between 10^{-3} and 10^{-4}, it decreases sharply after which a period of more predictable decrease is observed.
- Beyond this point, slower progress is made. We therefore choose our stopping tolerance to be $\varepsilon = 10^{-5}$ so that the algorithm will terminate in this region.

The change in successive iterates was found to also exhibit similar behavior (not shown), so is another suitable candidate for a stopping criterion.

It is worth noting that there are many other techniques for solving (variants of) the protein conformation problem (see for instance [22]). Such a discussion, however, is beyond the scope of this chapter.

5.7.1 Basic Douglas–Rachford Algorithm Results

Table 5.2 gives results for the basic Douglas–Rachford algorithm presented in Fig. 5.3. We make some comments regarding these results.

Table 5.2 Average (worst) results from five random replications of the basic Douglas–Rachford algorithm with $\varepsilon = 10^{-5}$

Protein	EDM-error		Position-error		Iterations		Time (h)	
1PTQ	3.6816	(4.0938)	0.1307	(0.1457)	4339.6	(4686)	0.28	(0.30)
1HOE	9.7475	(13.8503)	0.1781	(0.2636)	20794.4	(21776)	3.50	(3.67)
1LFB	9.8728	(17.2860)	1.1388	(2.1177)	22346.2	(23295)	4.64	(4.85)
1PHT	10.3709	(12.9557)	12.8782	(13.0056)	20103.0	(20251)	13.90	(14.00)
1POA	25.4225	(46.5804)	0.5844	(1.1639)	28426.0	(29766)	23.33	(24.47)
1AX8	25.7369	(39.4586)	0.6592	(0.9160)	17969.8	(19059)	15.04	(15.95)

The EDM-error increases with increasing problem size; yet the same trend is not observed for the position-error for which 1PHT reported the largest error. For all of the proteins studied, the differences between the average and worst-case results for the position-errors were small. This strongly suggests that the method can consistently produce a EDM which gives the desired atomic positions.

The second column of Fig. 5.6 shows the conformation of the basic Douglas–Rachford reconstructions, which are visually indistinguishable from the actual conformation shown in the first column. This is an improvement from what was reported in [3] whose Douglas–Rachford reconstructions of two of the larger proteins, 1POA and 1AX8, gave unrealistic conformations consisting of disjoint blocks of atoms. In light of Remark 5.6, it is likely that this was due to premature algorithm termination.

5.7.2 Douglas–Rachford Algorithm with Periodic Rank Projections

In our formulation of the protein confirmation problem, the most expensive step is the computation of the projection onto the rank constraint C_2 and thus requires the eigen-decomposition of a $(m - 1) \times (m - 1)$ symmetric matrix. In this section, we propose problem-specific heuristics which allow for this computation to sometimes be avoided.

One idea to avoid performing the eigen-decomposition is to not update the sequence $(r_n)_{n=1}^{\infty}$ in the algorithm given in Fig. 5.3 at every iteration but only periodically. This approach is described in Fig. 5.7 and results, with updates only every third time, in Table 5.3.

We now compare the results of this section to those of Sect. 5.7.1. A small increase in the position-errors and a larger increase in the EDM-errors were observed. The number of iterations required also increased, with this number almost doubling for 1PTQ. For all six test proteins, the total time required was less. The biggest improvement was 1POA whose total time was more than halved. The quality of the reconstructed conformations seems not to be adversely effected by the use of periodic rank projections, as shown in Fig. 5.6.

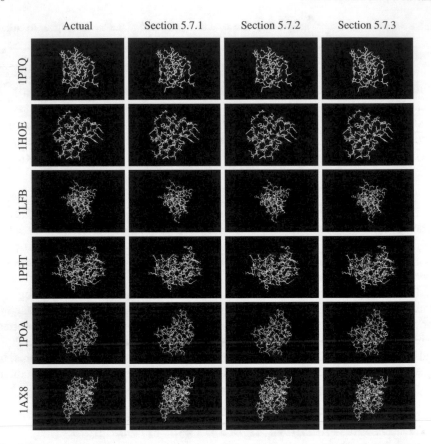

Fig. 5.6 The conformations of the six proteins and their three Douglas–Rachford reconstructions

Table 5.3 Average (worst) results from five random replications of the Douglas–Rachford algorithm with periodic rank projections with $T = 3$ and $\varepsilon = 10^{-5}$

Protein	EDM-error		Position-error		Iterations		Time (h)	
1PTQ	4.3709	(4.7200)	0.1919	(0.2240)	7160.6	(7595)	0.16	(0.17)
1HOE	10.1790	(12.1089)	0.2603	(0.2933)	20305.4	(22550)	1.21	(1.35)
1LFB	17.6532	(19.0984)	1.2709	(1.7243)	28983.8	(31211)	2.15	(2.31)
1PHT	23.8594	(25.9794)	13.1358	(13.2805)	20559.2	(20981)	5.03	(5.13)
1POA	49.8406	(51.3411)	1.0948	(1.2084)	33150.8	(39083)	9.55	(11.25)
1AX8	45.5203	(49.1866)	1.1696	(1.4482)	27080.6	(31250)	7.96	(9.20)

5.7.3 Reconstructions with Additional Distance Data

In Sects. 5.7.1 and 5.7.2, we considered the physically realistic setting in which distances below the threshold of 6 Å were known. As noted, when the number of

Input: $x_0 \in X, T \in \mathbb{N}$ and $\varepsilon > 0$
$n = 0$;
$p_0 \in P_A(x_0)$;
while $n = 0$ **or** $\|r_n - p_n\| > \varepsilon \|p_n\|$ **do**
    if $n \bmod T = 0$ **then**
       | $r_n \in P_B(2p_n - x_n)$;
    else
       | $r_n = r_{n-1}$;
    end
    $x_{n+1} = x_n + r_n - p_n$;
    $p_{n+1} \in P_A(x_{n+1})$;
    $n = n + 1$;
end
Output: $p_n \in X$

Fig. 5.7 The Douglas–Rachford algorithm with T-periodic projections onto the set B

Table 5.4 Average (worst) results from five random replications of the basic Douglas–Rachford algorithm from the smallest 10% of inter-atomic distances with $\varepsilon = 10^{-5}$

Protein	EDM-error		Position-error		Iterations		Time (h)	
1PTQ	3.1924	(3.5936)	0.0963	(0.1213)	4014.4	(4184)	0.26	(0.27)
1HOE	8.0905	(10.4357)	0.0960	(0.1265)	15110.4	(15709)	2.54	(2.64)
1LFB	7.2941	(13.9893)	0.4647	(0.9182)	11060.6	(11912)	2.29	(2.46)
1PHT	14.1302	(20.2476)	0.3542	(0.4326)	6071.0	(6512)	4.19	(4.49)
1POA	19.5619	(31.1987)	0.1624	(0.2665)	11555.8	(13244)	9.44	(10.81)
1AX8	14.0747	(29.7259)	0.0940	(0.1922)	10099.2	(11125)	8.38	(9.23)

atoms in a protein increases, the proportion of inter-atomic distances below this threshold compared to the total number of (nonzero) distances decreases.

To better understand the Douglas–Rachford method applied to larger problem instances, we performed the same reconstruction as in Sect. 5.7.1 but with the percentage of known nonzero distances constant. More precisely, we assumed that the smallest 10% of inter-atomic distances were known (Table 5.4).

As could perhaps be predicted, when more distance information is incorporated the error metrics and the number of iterations decrease. Problem size and EDM-error do not correlate as strongly compared to the results of Sect. 5.7.1. However, the general trend that larger problem sizes give larger EDM-errors is still observed. The most notable improvement, when compared to Sect. 5.7.1, is the position-error for 1PHT. This suggests that in the realistic setting of Sect. 5.7.1 the underlying protein's conformation (e.g., a compact or a dispersed conformation) is an important factor in the difficulty of the reconstruction problem.

5.7.4 Ionic Liquid Bulk Structure Determination

Ionic liquids (ILs) are salts (i.e., they are comprised of positively and negatively charged ions) having low melting points, typically occupying the liquid state at room temperature. An analogous reconstruction problem arising in the context of ionic liquid chemistry is to determine a given ionic liquid's *bulk structure*, that is, the configuration of its ions with respect to each other (the structure of the individual ions is known).

In this section, we applied the Douglas–Rachford method to a simplified version of this problem. Entries of the partial EDM are assumed to be known whenever the two atoms are bonded (i.e., when their *van der Waals radii* taken from [9] overlap).

Table 5.5 reports results for a *propylammonium nitrate (PAN)* data set consisting of 180 atoms. The corresponding rank-3 EDM completion problem has a total of 32,220 nonzero inter-atomic distances of which 5.95% form the partial EDM.

As was the case in the protein conformation application, the difference between the average and worst-case results for the two error metrics is observed to be small. The actual conformation of PAN and its Douglas–Rachford reconstruction are shown in Fig. 5.8. A high degree of visual coincidence is observed, although a small amount of the finer detail is missing.

Table 5.5 Average (worst) results from five random replications of the basic Douglas–Rachford algorithm, applied to ionic liquid bulk structure determination, with $\varepsilon = 10^{-5}$

EDM-error		Position-error		Iterations		Time (h)	
0.6323	(0.6918)	2.0374	(2.5039)	41553.2	(82062)	0.22	(0.43)

 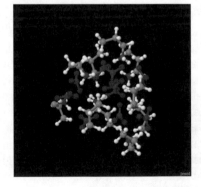

Fig. 5.8 The actual conformation (left) and Douglas–Rachford reconstruction (right) of PAN. Note the two poorly reconstructed hydrogen atoms (white) in the left configuration

5.8   Concluding Remarks

We have shown that the Douglas–Rachford reflection method can successfully solve the protein conformation determination problem by directly addressing a nonconvex matrix completion problem. This is also the case for an analogous ionic liquid bulk structure determination problem. It is worth emphasizing again that the current literature provides no theoretical justification for the method to work at all, let alone so well. Modifications of the method have also been shown to reduce computational times without significantly affecting the quality of the results. This promising demonstration of the method begs further attention, both in improving theoretical understanding and in the refinement and investigation of these and further applications.

Acknowledgements   The authors wish to thank Dr. Alister Page for introducing us to the bulk structure determination problem and for kindly sharing the PAN data set. The work of JMB is supported in part by the Australian Research Council. This work was performed during MKT's candidature at the University of Newcastle where he was supported in part by an Australian Postgraduate Award.

References

1. Aragón Artacho, F., Borwein, J.: Global convergence of a non-convex Douglas-Rachford iteration. J. Glob. Optim. **57**(3), 753–769 (2013)
2. Aragón Artacho, F., Borwein, J., Tam, M.: Recent results on Douglas–Rachford methods for combinatorial optimization problems. J. Optim. Theory Appl. (in press, 2013)
3. Aragón Artacho, F., Borwein, J., Tam, M.: Douglas-Rachford feasibility methods for matrix completion problems. ANZIAM J. **55**(4), 299–326 (2014)
4. Bauschke, H., Bello Cruz, J., Nghia, T., Phan, H., Wang, X.: The rate of linear convergence of the Douglas-Rachford algorithm for subspaces is the cosine of the Friedrichs angle. J. Approx. Theory **185**, 63–79 (2014)
5. Bauschke, H., Combettes, P.: Convex Analysis and Monotone Operator Theory in Hilbert Space. Springer, New York (2011)
6. Bauschke, H., Combettes, P., Luke, D.: Finding best approximation pairs relative to two closed convex sets in Hilbert spaces. J. Approx. Theory **127**(2), 178–192 (2004)
7. Bauschke, H., Noll, D., Phan, H.: Linear and strong convergence of algorithms involving averaged nonexpansive operators. J. Math. Anal. Appl. **421**(1), 1–20 (2015)
8. Berman, H., Westbrook, J., Feng, Z., Gilliland, G., Bhat, T., Weissig, H., Shindyalov, I.N., Bourne, P.E.: The protein data bank. Nucleic Acids Res. **28**, 235–242 (2000)
9. Bondi, A.: van der Waals volumes and radii. J. Phys. Chem. **68**(3), 441–51 (1964)
10. Borwein, J., Lewis, A.: Convex Analysis and Nonlinear Optimization. Springer, Berlin (2006)
11. Borwein, J., Sims, B.: The Douglas–Rachford algorithm in the absence of convexity. In: Fixed-Point Algorithms for Inverse Problems in Science and Engineering, pp. 93–109. Springer (2011)
12. Borwein, J., Tam, M.: The cyclic Douglas-Rachford method for inconsistent feasibility problems. J. Nonlinear Convex Anal. **16**(4), 537–584 (2015)
13. Borwein, J., Tam, M.: A cyclic Douglas-Rachford iteration scheme. J. Optim. Theory Appl. **160**(1), 1–29 (2014)
14. Borwein, J., Zhu, Q.: Techniques of Variational Analysis, *CMS Books in Mathematics*, vol. 20. Springer-Verlag, New York (2005, Paperback, 2010)

15. Berman, A., Shaked-Monderer, N.: Completely Positive Matrices. World Scientific, Singapore (2003)
16. Cegielski, A.: Iterative Methods for Fixed Point Problems in Hilbert Space. Lecture Notes in Mathematics, vol. 2057. Springer, London (2012)
17. Dattorro, J.: Convex Optimization & Euclidean Distance Geometry. Meboo Publishing USA (2005)
18. Elser, V., Rankenburg, I., Thibault, P.: Searching with iterated maps. Proc. Natl. Acad. Sci. **104**(2), 418–423 (2007)
19. Gravel, S., Elser, V.: Divide and concur: a general approach to constraint satisfaction. Phys. Rev. E **78**(3), 036706 (2008)
20. Hayden, T., Wells, J.: Approximation by matrices positive semidefinite on a subspace. Linear Algebra Appl. **109**, 115–130 (1988)
21. Hesse, R., Luke, D.: Nonconvex notions of regularity and convergence of fundamental algorithms for feasibility problems. SIAM J. Optim. **23**(4), 2397–2419 (2013)
22. Seo, J., Kim, J.-K., Ryu, J., Lavor, C., Mucherino, A., Kim, D.-S.: BetaMDGP: protein structure determination algorithm based on the Beta-complex. Trans. Comput. Sci. **8360**, 130–155 (2014)

Chapter 6
On Single-Valuedness of Quasimonotone Set-Valued Operators

Didier Aussel

Abstract A Nash problem is a noncooperative game in which the objective function of each player also depends on the decision variable of the other player. In order to solve such difficult problem, a classical approach is to write the optimality conditions of each of the problems obtaining thus a variational inequality. If the objective functions are nondifferentiable, the variational inequality can be set-valued, that is defined by a point-to-set map. Indeed, the derivatives are replaced by subdifferentials which are monotone if the objective functions are convex in the player's variable. And if the objective functions are quasiconvex in terms of the player's variable, the normal operator will advantageously replace the derivatives [3]. But solving a set-valued variational inequality is clearly more difficult than solving a single-valued variational inequality. It is thus very important to know sufficient conditions ensuring that a set-valued map, in particular a normal operator, is single-valued. Any monotone set-valued map that is also lower semi-continuous at a given point of the interior of its domain is actually single-valued at this point. This famous result is due to Kenderov [18] in 1975. Such a pointwise property of monotone maps has its local and dense counterparts (see, e.g. [14] and [10], respectively). The aim of this chapter is to answer to the somehow natural question: "what can one expect as a similar single-valuedness result for the more general class of quasimonotone set-valued maps". The three points of view, that is pointwise, local and dense aspects, are treated, and a central role is played by the concept of directional single-valuedness. For the sake of simplicity, in Sects. 6.3 and 6.4, we only consider finite dimensional setting even if all of the results of Sect. 6.3, respectively 6.4, hold true in Banach spaces, respectively in Hilbert spaces.

D. Aussel (✉)
Laboratory PROMES UPR CNRS 8521, University of Perpignan,
Perpignan, France
e-mail: aussel@univ-perp.fr

© Springer Nature Singapore Pte Ltd. 2017
D. Aussel and C. S. Lalitha (eds.), *Generalized Nash Equilibrium
Problems, Bilevel Programming and MPEC*, Forum for Interdisciplinary
Mathematics, https://doi.org/10.1007/978-981-10-4774-9_6

6.1   Introduction

A set-valued map $T : \mathbb{R}^n \rightrightarrows \mathbb{R}^n$ is said to be monotone on its domain Dom T if, for any (x, x^*) and (y, y^*) of its graph Gr T, one has $\langle y^* - x^*, y - x \rangle \geq 0$. One symbolic illustration of this is given by the subdifferential of the (convex) absolute value function $| \cdot | : x \mapsto |x|$. Indeed, the subdifferential $T := \partial | \cdot |$ of this simple function is described by

$$\partial | \cdot |(x) = \begin{cases} \{-1\} & \text{if } x < 0 \\ [-1, 1] & \text{if } x = 0 \\ \{1\} & \text{if } x > 0. \end{cases}$$

Nevertheless, one can immediately observe that this set-valued map $T = \partial | \cdot |$ actually satisfies that:

(i) it is single-valued at any x where the function $| \cdot |$ is continuously differentiable;
(ii) if T is single-valued at a point x, then it is so on a neighbourhood of x;
(iii) T is single-valued on a dense subset of Dom $T = \mathbb{R}$.

These observations are actually very particular cases of general and famous results that are cornerstone of the literature on monotone set-valued maps:

(i) [Kenderov (1975)] [18]: Let $T : \mathbb{R}^n \rightrightarrows \mathbb{R}^n$ be a monotone set-valued map. If T is lower semi-continuous at $x_0 \in \text{int}(\text{Dom } T)$, then T is single-valued at x_0;
(ii) [Dontchev-Hager (1994)] [14] Let $T : \mathbb{R}^n \rightrightarrows \mathbb{R}^n$ be a monotone set-valued mapping and $x_0 \in \text{int}(\text{Dom } T)$. Assume that T has the Aubin property around $(x_0, x_0^*) \in \text{Gr } T$, then T is locally single-valued at x_0;
(iii) [Borwein-Lewis (2006)] [10]: any minimal cusco map (convex compact-valued upper semi-continuous map)—this covers as particular case maximal monotone maps—is single-valued on a dense set of its domain.

Our aim in the chapter is to focus on the widely more general class of quasimonotone set-valued maps and to see if some equivalent to the three results above can be proved for this class. Let us recall that a set-valued map $T : \mathbb{R}^n \rightrightarrows \mathbb{R}^n$ is said to be *quasimonotone* if, for all $(x, x^*) \in \text{Gr } (T)$ and any $(y, y^*) \in \text{Gr } (T)$, one has $\langle x^*, y - x \rangle > 0 \Rightarrow \langle y^*, y - x \rangle \geq 0$.

It is nevertheless clear that this naive idea cannot resist to a first and rapid analysis. Indeed, if one simply considers the set-valued map

$$T : \mathbb{R} \rightrightarrows \mathbb{R}$$
$$x \mapsto T(x) = \mathbb{R}^{++}$$

then clearly T is quasimonotone (but not monotone), satisfies the Aubin property around any $(x, x^*) \in \text{Gr } T$, is lower semi-continuous at any $x \in \text{Dom } T = \mathbb{R}$ but it is set-valued at any point. It simply means that the concept of single-valuedness is not adapted to the case of quasimonotone operator. Indeed, by definition the quasi-monotonicity is a "directional concept", and thus, taking this intrinsic characteristic into account, in [6], the authors proposed the new notion of *single-directionality*.

The chapter is organized as follows. First, the basic necessary notions, in particular the single-directionality, are recalled in Sect. 6.2. We have seen above that the single-valuedness results for monotone maps can be classified into three categories: pointwise, local or dense points of view. Following the same classification, Sects. 6.3, 6.4 and 6.5, respectively, cover these three points of view for the single-directionality of quasimonotone maps. For each of these cases, an application to the *normal operator* of quasiconvex functions, recalled in Sect. 6.2.3, is considered.

Let us note that the above three fundamental results of monotone set-valued maps have been stated in infinite dimensional spaces: Banach spaces in Kenderov [18] and Dontchev-Hager [14]. This chapter is essentially based on [4–6, 8] where the results are also established in general Banach spaces. Nevertheless, for the sake of simplicity, in Sects. 6.3 and 6.4 of this chapter we have chosen to restrict ourself to the finite dimension case.

6.2 Notations

6.2.1 Basic Definitions

Throughout the paper, for any $x \in \mathbb{R}^n$ and $\rho > 0$, we denote by $B(x, \rho)$, $\overline{B}(x, \rho)$ and $S(x, \rho)$, respectively, the open ball, the closed ball and the sphere of center x and radius ρ, while for $x, x' \in \mathbb{R}^n$, we denote by $[x, x']$ the closed segment $\{tx + (1 - t)x' : t \in [0, 1]\}$. The segments $]x, x'[$, $]x, x']$, $[x, x'[$ are defined analogously. For any element x^* of $\mathbb{R}^n$, we set $\mathbb{R}_+\{x^*\} = \{tx^* \in \mathbb{R}^n : t \geq 0\}$ and $\mathbb{R}_{++}\{x^*\} = \{tx^* \in \mathbb{R}^n : t > 0\}$.

The topological closure, the interior, the boundary and the convex hull of a set $A \subset \mathbb{R}^n$ will be denoted, respectively, by $cl(A)$, $int(A)$, $bd(A)$ and $conv(A)$. Given any nonempty subset A of $\mathbb{R}^n$ and a point $y \in \mathbb{R}^n$, the distance from y to A will be denoted by $dist(y, A) = \inf\{\|y - y'\| : y' \in A\}$ and $N(A, y)$ stands for the normal cone to A at y, that is

$$N(A, y) = \{y^* \in \mathbb{R}^n : \langle y^*, u - y \rangle \leq 0, \ \forall u \in A\}.$$

The domain and the graph of a set-valued operator $T : \mathbb{R}^n \rightrightarrows \mathbb{R}^n$ will be denoted, respectively, by Dom T and Gr T, while the inverse image of T at x will be $T^{-1}(x) = \{y \in \mathbb{R}^n : x \in T(y)\}$. The map T is said to be *trivial at* $x \in$ Dom T if $T(x) = \{0\}$.

Let us recall the classical definitions of generalized monotonicity of set valued maps that will be used in the sequel. A set-valued map $T : \mathbb{R}^n \rightrightarrows \mathbb{R}^n$ is said to be

– *quasimonotone* if, for all $(x, x^*) \in$ Gr (T) and any $(y, y^*) \in$ Gr (T),

$$\langle x^*, y - x \rangle > 0 \implies \langle y^*, y - x \rangle \geq 0.$$

– *pseudomonotone* if, for all $(x, x^*) \in \mathrm{Gr}\,(T)$ and any $(y, y^*) \in \mathrm{Gr}\,(T)$,

$$\langle x^*, y - x \rangle \geq 0 \;\Rightarrow\; \langle y^*, y - x \rangle \geq 0.$$

Clearly, pseudomonotonicity implies quasimonotonicity. These concepts of generalized monotonicity play a fundamental role in the first-order analysis of optimization problem with quasiconvex or pseudoconvex objective functions (see, e.g. Aussel-Hadjisavvas [7]).

An interesting link between the monotonicity of a set-valued map and the quasimonotonicity of its linear perturbations has been remarked in Aussel-Lassonde-Corvellec [1].

Proposition 6.1 [1] *Let* $T : \mathbb{R}^n \rightrightarrows \mathbb{R}^n$ *be a set-valued map. Then, T is monotone if and only if $T + \alpha^*$ is quasimonotone, for all $\alpha^* \in \mathbb{R}^n$.*

On the other hand, we will use the following continuity notions for a set-valued operator $T : \mathbb{R}^n \rightrightarrows \mathbb{R}^n : T$ is said to be

– *lower semi-continuous* at $x \in \mathrm{Dom}\,T$ if, for every open set V intersecting $T(x)$, there is a neighbourhood U of x such that $V \cap T(x') \neq \emptyset$, for all $x' \in U$.
– *upper semi-continuous* at $x \in \mathrm{Dom}\,T$ if, for every open set W containing $T(x)$, there is a neighbourhood U of x such that $W \supset T(x')$ for all $x' \in U$.
– *lower sign-continuous* at $x \in \mathrm{Dom}\;T$ [17] if, for any $v \in \mathbb{R}^n$, the following implication holds:

$$\left(\forall t \in \;]0, 1[, \;\; \inf_{x_t^* \in T(x_t)} \langle x_t^*, v \rangle \geq 0 \right) \;\Rightarrow\; \inf_{x^* \in T(x)} \langle x^*, v \rangle \geq 0$$

where $x_t = x + tv$.
– *upper sign-continuous* at $x \in \mathrm{Dom}\;T$ [17] if, for all $v \in \mathbb{R}^n$, the following implication holds:

$$\forall t \in \;]0, 1[, \;\; \inf_{x_t^* \in T(x + tv)} \langle x_t^*, v \rangle \geq 0 \Rightarrow \sup_{x^* \in T(x)} \langle x^*, v \rangle \geq 0.$$

It follows from the definitions that any lower sign-continuous map is upper sign-continuous. Roughly speaking, the lower sign-continuity corresponds to a "directional version" of the lower semi-continuity. Indeed, as one can observe from the definition, this property does not depend on the norm of the nonzero elements of the values $T(x)$ but on their direction.

Clearly, any lower semi-continuous operator is lower sign-continuous. Let us describe, in the following exercise, a simple example of a lower sign-continuous map which is not lower semi-continuous.

Exercise 6.1 Our aim is to prove that the Clarke subdifferential of any composition $h = g \circ f$ of a real-valued continuously differentiable function f, defined on $\mathbb{R}^n$, with a locally Lipschitz function g such that, for any $t \in \mathbb{R}$, $\partial g(t) \subset \mathbb{R}_+^*$ is lower sign-continuous at any point x_0 such that $\nabla f(x_0) \neq 0$. The proof can be done following the following steps:

(1) Recall that such a subdifferential ∂h is known, in general, to be upper semi-continuous but not lower semi-continuous. Give an example of such a composition h for which the subdifferential not lower semi-continuous.
(2) Prove that there exists a neighbourhood U of x_0, such that

$$\partial h(u) = \partial g(f(u)).\{\nabla f(u)\}, \quad \forall u \in U. \tag{6.1}$$

Deduce that, for any $u \in U$, one has $\mathbb{R}_+^* \partial h(u) = \mathbb{R}_+^* \{\nabla f(u)\}$.
(3) Assume that v is any elements of $\mathbb{R}^n$ such that for any $t \in]0, 1[$,

$$\inf_{x_t^* \in \partial h(x_t)} \langle x_t^*, v \rangle \geq 0$$

where x_t stands for $x_t = x_0 + tv$. Show that $\langle \nabla f(x_0), v \rangle \geq 0$.
(4) Conclude to the lower sign-continuity of h at x_0.

Other examples showing the difference between both concepts of semi-continuity can be found in Aussel-Cotrina [2]. For example, it is shown in this work that a set-valued operator T is lower sign-continuous at a given point if and only if the convexified version of T, that is the operator having as values the convex hull of the values of T, is lower sign-continuous. This property is clearly not satisfied for lower semi-continuity.

It is important to notice that if an operator T is lower semi-continuous at a point x of its domain, then x is an element of int(Dom T). This comes directly from the definition of lower semi-continuity. On the contrary, as can be seen in the following example, a map can be lower sign-continuous on the boundary of its domain.

Example 6.1 Let Z be a closed proper subspace of $\mathbb{R}^n$, and let us denote by i_Z the indicator function of Z, that is $i_Z(z) = 0$ if $z \in Z$ and $i_Z(z) = +\infty$ if not. If T stands for the subdifferential of i_Z, that is

$$T(z) = \begin{cases} \partial i_Z(z) = Z^\perp & \text{if } z \in Z, \\ \partial i_Z(z) = \emptyset & \text{otherwise} \end{cases}$$

then, clearly, one has int(Dom T) $= \emptyset$, and thus, T is nowhere lower semi-continuous. But, it is easy to see that T is lower sign continuous everywhere on Dom $T = Z$.

The general interrelations between those different concepts of continuity are summarized in the following figure:

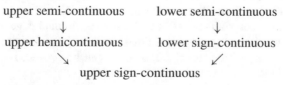

Finally, the following concepts of localization of a set-valued map will be used in the sequel. Given a set-valued map $T : \mathbb{R}^n \rightrightarrows \mathbb{R}^n$, an open set U of $\mathbb{R}^n$ and an open set V of $\mathbb{R}^n$:

- the operator $T_{U \times V}$ defined by $\mathrm{Gr}\,(T_{U \times V}) = \mathrm{Gr}\,(T) \cap (U \times V)$ is called a *localization* of T (Levy-Poliquin [21]);
- the operator $T_{U \times V}$ is said to be a *localization of T at x* if $T_{U \times V}$ is a localization of T, x is an element of U and $T(x) \cap V$ is nonempty.

6.2.2   Single-Directionality

As explained in the introduction, the concept of single-valuedness is not adapted to the analysis of quasimonotone operators, since this generalized monotonicity is intrinsically of directional nature. This is why in [6], the notion of *single-directionality* of a set-valued map has been introduced.

Definition 6.1 A set-valued operator $T : \mathbb{R}^n \rightrightarrows \mathbb{R}^n$ is said to be

- *single-directional* at $y \in \mathrm{Dom}\ T$ if $T(y) \subseteq \mathbb{R}_+\{y^*\}$ for some $y^* \in T(y)$,
- *locally single-directional* at $y \in \mathrm{Dom}\ T$ if there exists a neighbourhood U of y such that for all $y' \in U$, $T(y')$ is single-directional.

Finally, T is said to be *strictly single-directional* (respectively, *locally strictly single-directional*) at y if $T(y) \subseteq \mathbb{R}_{++}\{y^*\}$ for some $y^* \neq 0$ (respectively, T is strictly single-directional at any point of some neighbourhood of y).

This single-directional property of a map T can be rephrased as a single-valuedness property of a normalized version T_α of T. Indeed, for any $\alpha > 0$, we can define the *normalized operator T_α, given for each $x \in \mathrm{Dom}\ T_\alpha = \mathrm{Dom}\ T$ by*

$$T_\alpha(x) = \begin{cases} \{0\} & \text{if}\quad T(x) = \{0\} \\ \left\{ \dfrac{1}{\alpha \|x^*\|} x^* \ :\ x^* \in T(x) \setminus \{0\} \right\} & \text{otherwise.} \end{cases}$$

From this definition, one immediately has, for any $\alpha > 0$ and any $x \in \mathrm{Dom}\ T$,

$$T \text{ is single-directional at } x \iff T_\alpha \text{ is single-valued at } x.$$

As observed in Aussel-Dutta [3, Lemma 3.1], one of the advantages of this normalization is that T_α is always bounded-valued and that the Stampacchia variational inequalities defined by T and T_α share the same nontrivial solutions, a trivial solution being a point of the constraint set such that 0 belongs to the image of the point.

On the other hand, if a map is single-valued at a point, it is clearly single-directional at that point. A kind of reciprocal implication can be proved, again using linear perturbations of the operator.

Lemma 6.1 *Let $T : \mathbb{R}^n \rightrightarrows \mathbb{R}^n$ be a set-valued map, x be a point of its domain* Dom T *and $x^* \in T(x)$. The map $T + \{\alpha^*\}$ is single-directional at x, for any $\alpha^* \in \mathbb{R}^n \setminus \{-x^*\}$, if and only if T is single-valued at x.*

Proof It is clear that if T is single-valued at x, then $T + \{\alpha^*\}$ is single-directional at x. Conversely, assume that there exists $y^* \in T(x)$ with $y^* - x^* \neq 0$. For $\alpha^* = -\frac{1}{2}(y^* - x^*) - x^* \neq -x^*$, there exists $w^* \in \mathbb{R}^n \setminus \{0\}$ such that $T(x) + \{\alpha^*\} \subset \mathbb{R}^+\{w^*\}$. In particular, one have

$$-\frac{1}{2}(y^* - x^*) = x^* + \alpha^* \in \mathbb{R}^+\{w^*\} \quad \text{and} \quad \frac{1}{2}(y^* - x^*) = y^* + \alpha^* \in \mathbb{R}^+\{w^*\}$$

which is impossible because $y^* - x^* \neq 0$.

6.2.3   Normal Operator in Quasiconvex Optimization

In what follows we shall deal with *proper* functions $g : \mathbb{R}^n \to \mathbb{R} \cup \{+\infty\}$ (i.e. functions for which dom $g = \{y : g(y) < +\infty\}$ is nonempty), and we will consider some generalized convexity assumptions over them. So let us recall that a function $g : \mathbb{R}^n \to \mathbb{R} \cup \{+\infty\}$ is said to be

- *quasiconvex* on a subset $C \subset$ dom g if, for any $y, y' \in C$ and any $t \in [0, 1]$,

$$g(ty + (1 - t)y') \leq \max\{g(y), g(y')\},$$

- *semi-strictly quasiconvex* on a subset $C \subset$ dom g if, g is quasiconvex and for any $y, y' \in K$,

$$g(y) < g(y') \Rightarrow g(z) < g(y') \quad \forall z \in [y, y'[.$$

Let us denote, for any $\alpha \in \mathbb{R}$, by $S_\alpha(g)$ and $S_\alpha^<(g)$ the sublevel set and the strict sublevel set, respectively, associated with g and α :

$$S_\alpha(g) = \{y \in \mathbb{R}^n : g(y) \leq \alpha\} \quad \text{and} \quad S_\alpha^<(g) = \{y \in \mathbb{R}^n : g(y) < \alpha\}.$$

Whenever no confusion can occur, we will use, for any $y \in$ dom g, the simplified notation $S_{g(y)}$ and $S_{g(y)}^<$ instead of $S_{g(y)}(g)$ and $S_{g(y)}^<(g)$. It is well known that the quasiconvexity of a function g is characterized by the convexity of the sublevel sets (or the convexity of the strict sublevel sets). Analogously, it is easy to check that any lower semi-continuous function g, semi-strictly quasiconvex on its domain dom g, satisfies the following property:

$$\forall \alpha > \inf_{\mathbb{R}^n} g, \quad \text{cl}\left(S_\alpha^<(g)\right) = S_\alpha(g). \tag{6.2}$$

Roughly speaking, this means that a lower semi-continuous semi-strictly quasiconvex function g does not have any "flat part" with nonempty interior on
dom $g \setminus \mathrm{argmin}_{\mathbb{R}^n} g$.

Simple examples of quasiconvex functions are "increasing step functions" of
real variable; thus showing that, by nature, those functions could be nonsmooth
(i.e. nondifferentiable). A classical tool for the study of first-order properties of
nonsmooth functions is the subdifferential, with various possible definitions. But all
of them are based on the construction of the normal cone to the epigraph of the
function, while, for quasiconvex function, the interesting properties come from the
geometry of the sublevel sets, and particularly from their convexity.

This is why some authors [11, 16] developed the so-called *normal operator*
defined, roughly speaking, as the normal cone to the sublevel sets. Nevertheless,
the class of quasiconvex functions also includes functions with "full dimension flat
parts" for which the normal cones to the sublevel sets $S_{g(x)}^{\leq}$ and $S_g^{<}(x)$ may fail to
provide strong properties as quasimonotonicity and/or upper semi-continuity. This
is the reason why more elaborated concepts of sublevel set and normal operator have
been defined in [7]: the *adjusted sublevel set* and its associated *adjusted normal
operator* are:

$$S_g^a(x) = \begin{cases} S_{g(x)}^{\leq} \cap \overline{B}\big(S_{g(x)}^{<}, \rho_x\big) & \text{if } x \notin \mathrm{argmin}\, g \\ S_{g(x)}^{\leq} & \text{otherwise} \end{cases}$$

where $\rho_x = \mathrm{dist}\big(x, S_{g(x)}^{<}\big)$ and

$$N_g^a(x) = \big\{x^* \in \mathbb{R}^n : \langle x^*, y - x\rangle \leq 0 \ \text{ for all } \ y \in S_g^a(x)\big\}.$$

In order to illustrate the definitions, let us consider a simple function g defined on
$[-5, 5]^2$ by

$$g(x) = \begin{cases} max\{x_1, x_2\} & \text{if } max\{x_1, x_2\} \leq 0 \\ \|(x_1 + 5, x_2 + 5)\|_2 - 8 & \text{if } \|(x_1 + 5, x_2 + 5)\|_2 \geq 8 \\ 0 & \text{otherwise} \end{cases} \qquad (6.3)$$

The following figures show the graph and the sublevel sets of g.

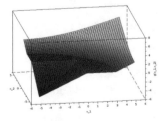
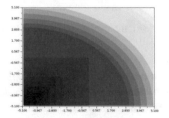

Then, the adjusted sublevel set at the point x and its normal operator are represented in the following figures:

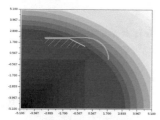 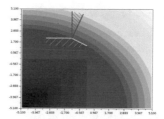

Let us notice that since $S^<_{g(y)} \subset S^a_g(y) \subset S_{g(y)}$, one has

$$N(S_{g(y)}, y) \subset N^a_g(y) \subset N(S^<_{g(y)}, y) \quad \forall y \in \text{dom } g. \tag{6.4}$$

Note that in the case of a semi-strictly quasiconvex function f, the three cones coincide. Imitating [4], we say that a function $g : \mathbb{R}^n \longrightarrow \mathbb{R} \cup \{+\infty\}$ is *solid* if, for any $x \notin \arg\min_{\mathbb{R}^n} g$, the sublevel set $S^<_{g(x)}$ has an interior point. Many precious properties of the adjusted normal operator have been proved for quasiconvex functions (see [7, 9]).

Proposition 6.2 *Let $g : \mathbb{R}^n \longrightarrow \mathbb{R} \cup \{+\infty\}$ be any function. Then,*

(i) *The mapping $x \longmapsto N^a_g(x)$ is quasimonotone.*
(ii) *If g is quasiconvex solid and lower semi-continuous, then $N^a_g(x) \neq \{0\}$ for every $x \in \text{Dom } g \setminus \arg\min_{\mathbb{R}^n} g$.*

Since the set-valued map N^a is cone-valued, there is no hope to obtain an upper semi-continuity of N^a. But an adapted concept is called the *cone upper semi-continuity*: An operator $T : \mathbb{R}^n \rightrightarrows \mathbb{R}^n$ whose values are convex cones is called *cone upper semi-continuous* at $x \in \text{dom } (T)$ if there exists a neighbourhood U of x and a base $C(u)$ of $T(u)$ for each $u \in U$, such that $u \rightarrow C(u)$ is upper semi-continuous at x. Combining Proposition 2.2 and Proposition 3.5 of [7], one can obtain the cone upper semi-continuity of the adjusted normal operator under mild assumptions.

Proposition 6.3 *If $g : \mathbb{R}^n \longrightarrow \mathbb{R} \cup \{+\infty\}$ is a quasiconvex and solid function, then at any point of $\text{Dom } (g) \setminus \text{argmin}_{\mathbb{R}^n} g$ where g is lower semi-continuous, its associated normal operator N^a_g is cone upper semi-continuous.*

6.3 Pointwise Single-Directionality

Let us now prove, in a very elementary way, our first single-directional result for quasimonotone set-valued maps. This theorem will play a central role in this chapter.

Theorem 6.1 *Let* $T : \mathbb{R}^n \rightrightarrows \mathbb{R}^n$ *be a quasimonotone set-valued map and* x *be an element of* int(Dom T). *If one of the following assumptions holds:*

(i) *T is lower sign-continuous at x;*
(ii) *There exists a localization* $T_{U \times V}$ *of T at x which is lower sign-continuous at x and nontrivial at x (i.e.* $T_{U \times V}(x) \neq \{0\}$);

then T is single-directional at x.

The result of Theorem 3.2 (i) was envisioned by Hadjisavvas [17], under Proposition 3.9, and proved in that proposition for the pseudomonotone case. Except for the localization issue, the proof of Theorem 3.2 is very similar to that in [17].

Proof Let us prove first case (*ii*). Let $\varepsilon > 0$ be such that $B(x, \varepsilon) \subset [\text{Dom } T \cap U]$. We will show that all the nonzero elements of $T(x)$ follow the same direction. Since $T_{U \times V}$ is nontrivial, let x^* denote an element of $T_{U \times V}(x) \setminus \{0\}$. If $T(x)$ does not contain any other nonzero element than x^*, then there is nothing to prove. Otherwise, let y^* be a point of $T(x) \setminus \{0, x^*\}$. Assume, for a contradiction, that $y^* \notin \mathbb{R}_{++}\{x^*\}$, which immediately implies that there exists $d \in B(0, 1)$ such that

$$\langle x^*, d \rangle < 0 < \langle y^*, d \rangle. \tag{6.5}$$

Since, for all $t \in]0, 1[, \langle y^*, (x + t\varepsilon d) - x \rangle = t\varepsilon \langle y^*, d \rangle > 0$, by quasimonotonicity of T, it follows that $\langle x_t^*, (x + t\varepsilon d) - x \rangle \geq 0$, for any $x_t^* \in T(x + t\varepsilon d)$. Thus,

$$\forall t \in]0, 1[, \quad \inf_{x_t^* \in T(x+t\varepsilon d)} \langle x_t^*, \varepsilon d \rangle \geq 0$$

and, according to the lower sign-continuity of $T_{U \times V}$, one has

$$\inf_{z^* \in T_{U \times V}(x)} \langle z^*, \varepsilon d \rangle \geq 0,$$

which contradicts (6.5).

Now case (*i*) is a direct consequence of case (*ii*) by considering the neighbourhoods $U = \mathbb{R}^n$ and $V = \mathbb{R}^n$.

Remark 6.1 (*a*) The following example shows that, in Theorem 6.1, we cannot replace the lower sign-continuity by the upper sign-continuity even if T is compact-valued. Let $T : \mathbb{R} \rightrightarrows \mathbb{R}$ given by

$$T(x) = \begin{cases} -1, & \text{if } x < 0, \\ [-1, 1], & \text{if } x = 0, \\ 1, & \text{if } x > 0. \end{cases}$$

This (quasi)monotone operator is upper sign-continuous at 0 but obviously not single-directional at 0.

(*b*) Looking to the proof of Theorem 6.1, one can actually wonder if, for a quasimonotone operator, lower sign-continuity can characterize single-valuedness. A slight modification of the above example, that is just considering $T(x) = -1$ if $x \leq 0$ and $T(x) = 1$ otherwise, shows that it is not the case. Indeed, T is single-valued and thus single-directional but it is not lower sign-continuous.

(*c*) Let us now observe that, even if an operator T is not single-directional at a point x, it may be possible to find a localization $T_{U \times V}$ of T at x which is single-directional, not only at x but at any point of U. This is, for example, the case for the following operator defined in [6]: let $T : \mathbb{R}^2 \rightrightarrows \mathbb{R}^2$ be such that

$$T(x) = \begin{cases} \{x\}, & \text{if } x \neq (0,0), \\ \{(0,0)\} \cup S((0,0),1), & \text{if } x = (0,0). \end{cases}$$

This pseudomonotone map is neither single-directional at $(0,0)$ nor lower sign-continuous at $(0,0)$. But, on the other hand, the localization $T_{U \times V}$ of T, with $U = V =]-1/4, 1/4[\times]-1/4, 1/4[$, is lower sign-continuous at any point of U, and thus, thanks to Theorem 6.1, it is single-directional at any point of U. Actually, as proved in the above theorem, if the localization $T_{U \times V}$ is not trivial at $x \in U \cap \text{int}(\text{Dom } T)$, then it is the entire operator T which will be proved to be single-directional at x.

As an immediate consequence of Theorem 6.1, we have the following statement which provides a "quasimonotone/single-directional" counterpart of the Kenderov's single-valuedness result.

Corollary 6.1 *Let $T : \mathbb{R}^n \rightrightarrows \mathbb{R}^n$ be a quasimonotone set-valued map. If T is lower semi-continuous at $x \in \text{int}(\text{Dom } T)$, then T is single-directional at x.*

Now combining Theorem 6.1, Proposition 6.1 and Lemma 6.1, we obtain, in a very simple way, the famous single-valuedness result of Kenderov.

Corollary 6.2 (Kenderov [18]) *Let $T : \mathbb{R}^n \rightrightarrows \mathbb{R}^n$ be a monotone set-valued map. If T is lower semi-continuous at $x_0 \in \text{int}(\text{Dom } T)$, then T is single-valued at x_0.*

Proof According to Proposition 6.1, the map $T + \alpha^*$ is quasimonotone and lower semi-continuous at x_0, for any $\alpha^* \in \mathbb{R}^n$. Thus by Theorem 6.1, $T + \alpha^*$ is single-directional at x_0 and the proof is complete by invoking Lemma 6.1.

It is important to notice that there is no hope to obtain, in the monotone case, a single-valuedness result under a regularity hypothesis as weak as in Theorem 6.1, that is only assuming the lower sign-continuity at the considered point. This can be seen on the following operator $T : \mathbb{R} \rightrightarrows \mathbb{R}$ given by

$$T(x) = \begin{cases} \{1\}, & \text{if } x < 0, \\ [1,2], & \text{if } x = 0, \\ \{2\}, & \text{if } x > 0, \end{cases}$$

which is clearly lower sign-continuous at $x_0 = 0$. The interested reader can have a look to Kenderov [19], Christensen [12], Christensen and Kenderov [13] and

Kenderov-Moors-Revalski [20] for pointwise single-valuedness results for monotone map under weaker assumptions than lower semi-continuity.

Now, if the map T is not only quasimonotone but pseudomonotone, then some
strict single-directionality can be deduced.

Theorem 6.2 *Let $T : \mathbb{R}^n \rightrightarrows \mathbb{R}^n$ be a pseudomonotone set-valued map and x be an
element of* int(Dom T). *If T admits a localization $T_{U \times V}$ at x which is lower sign-
continuous at x and nontrivial at x, then T is strictly single-directional at x.*

Proof Let us suppose that T is not trivial at x. According to Theorem 6.1, T is
single-directional at x. So if $0 \notin T(x)$, the latter immediately means that T is in fact
strictly single-directional at x. Thus, assume now that $0 \in T(x)$. Let $\varepsilon > 0$ be such
that $B(x, \varepsilon) \subset [\text{Dom } T \cap U]$. By the pseudomonotonicity of T, for any $d \in X$ with
$\|d\| = 1$, we have $\langle x_t^*, \varepsilon d \rangle \geq 0$, for any $x_t^* \in T_{U \times V}(x + t\varepsilon d)$. Hence,

$$\inf_{x_t^* \in T_{U \times V}(x + t\varepsilon d)} \langle x_t^*, \varepsilon d \rangle \geq 0,$$

thus, again by lower sign-continuity of $T_{U \times V}$ at x,

$$\inf_{x^* \in T_{U \times V}(x)} \langle x^*, \varepsilon d \rangle \geq 0.$$

Since the last inequality holds true for any direction $d \in B(0, 1)$, it follows that
$T_{U \times V}(x) = \{0\}$ and the proof is complete.

Let us observe that the strict single-directional property cannot be reached, in general,
if T is not pseudomonotone. This is, for example, the case for the map $T : \mathbb{R} \rightrightarrows \mathbb{R}$
defined by $T(x) = \mathbb{R}^+$ which is, nevertheless, quasimonotone, lower sign-continuous
and nontrivial on $\mathbb{R}$.

It is important to notice that the above results (Theorems 6.1 and 6.2) cannot be
extended to the entire domain of the operator. This can be seen by considering the
map proposed in Example 6.1. Indeed even though the map T defined in this example
is maximal monotone and is lower sign-continuous everywhere in Dom $T = Z$, it is
nowhere single-directional as soon as the dimension of the orthogonal space $Z^\perp$ of
Z is strictly greater than one.

To end this section, let us illustrate, on the quasiconvex function g defined by
formula (6.3), how the above results can apply to the quasimonotone adjusted normal
operator. From the definition of the adjusted normal operator N_g^a is lower semi-
continuous at any point of $\mathscr{C} =] - 5, 5[^2 \backslash \mathscr{D}$ where

$$\mathscr{D} = [([-5, 0] \times \{0\}) \cup (1 \times [-5, 0]) \cup \left\{ (x_1, x_2) : \begin{array}{l} \|(x_1 + 5, x_2 + 5)\|_2 = 8\} \\ \text{or} \\ x_1 \leq 0 \text{ and } x_2 \leq 0 \text{ and } x_1 \neq x_2 \end{array} \right\}$$

Thus according to Theorem 6.1, the normal operator N_g^a is single-directional at each
point of $\mathscr{C}$.

6.4 Local Single-Directionality

In Sect. 6.4.2, we have considered the pointwise approach. Nevertheless, there exist also in the literature many results concerning local behavior, for example Donchev-Hager [14] and Levy-Poliquin [21] for the monotone case and Aussel-Garcia-Hadjisavvas [6] for the quasimonotone case. In this section, we show that actually those local type results can be deduced from their pointwise counterparts.

6.4.1 General Results

Recall that a set-valued operator $T : \mathbb{R}^n \rightrightarrows \mathbb{R}^n$ is said to have the *Aubin property around* $(x, x^*) \in \operatorname{Gr} T$ if there exist neighbourhoods U of x, V of x^* and a positive real number l such that

$$T(u) \cap V \subset T(u') + \bar{B}_{\mathbb{R}^n}(0, l\|u' - u\|), \quad \forall u, u' \in U. \tag{6.6}$$

The following lemma enlightens the fact that under Aubin property, the pointwise nontriviality immediately extends to a local nontriviality.

Lemma 6.2 *Suppose that* $T : \mathbb{R}^n \rightrightarrows \mathbb{R}^n$ *has the* Aubin property around $(x_0, x_0^*) \in \operatorname{Gr} T$ *(with respectively neighbourhoods U and V). If the localization $T_{U \times V}$ is nontrivial at x_0, then there exists $\varepsilon > 0$ such that $T_{U \times V}$ is nontrivial on $B(x_0, \varepsilon)$, where $B(x_0, \varepsilon) \subset U$.*

Proof Take $\varepsilon > 0$ such that $l\varepsilon < \|x_0^*\|$, $B(x_0, \varepsilon) \subset U$ and $B(x_0^*, l\varepsilon) \subset V$ where l is the positive real number in the Aubin property. Let $x \in B(x_0, \varepsilon)$. From (6.6), there exists $x^* \in T(x)$ such that

$$x_0^* = x^* + l\|x - x_0\|e, \quad \text{with } \|e\| \leq 1.$$

Hence, $\|x_0^* - x^*\| \leq l\|x - x_0\| \leq l\varepsilon < \|x_0^*\|$, and it follows that $x^* \neq 0$ with $x^* \in T(x) \cap B(x_0^*, l\varepsilon) \subset T(x) \cap V$.

The following interesting result, that first appears in Levy-Poliquin [21, Lemma 2.3], states a link between the Aubin property of the operator and the lower semicontinuity of one of its localization.

Lemma 6.3 ([21]) *If the set-valued map* $T : \mathbb{R}^n \rightrightarrows \mathbb{R}^n$ *has the Aubin property around $(x_0, x_0^*) \in \operatorname{Gr} T$ (with U and V), then the localization $T_{U \times V}$ at x_0 is lower semi-continuous on U.*

We are now in position to establish, combining Theorem 6.1 and Lemma 6.3, several local single-directionality results under generalized monotonicity assumptions.

Proposition 6.4 *Let $T : \mathbb{R}^n \rightrightarrows \mathbb{R}^n$ be a set-valued map and $x_0 \in \text{Dom } T$. Assume that T has the Aubin property around $(x_0, x_0^*) \in \text{Gr } T$ (with U and V). Then,*

(i) If T is quasimonotone, then the localization $T_{U \times V}$ is single-directional on U.

(ii) If T is quasimonotone and $T_{U \times V}$ is nontrivial at x_0, then T itself is locally single-directional at x_0.

(iii) If T is pseudomonotone and $T_{U \times V}$ is nontrivial at x_0, then T is locally strictly single-directional at x_0.

Proof First of all, from Lemma 6.3, the localization $T_{U \times V}$ is lower semi-continuous and thus lower sign-continuous, on U. Therefore, applying Theorem 6.1 (*i*) to $T_{U \times V}$, we obtain assertion (*i*). If, additionally, the localization $T_{U \times V}$ is nontrivial at x_0, then by Lemma 6.2, there exists $\varepsilon > 0$ such that $T_{U \times V}$ is nontrivial on $B(x_0, \varepsilon)$ and thus $T_{B(x_0, \varepsilon) \times V}$ is single-directional on $B(x_0, \varepsilon)$, according to Theorem 6.1 (*ii*). Finally assertion (*iii*) can be obtained, using Theorem 6.2, by the same arguments.

Finally, using Lemma 6.3, the following "localization version" of the single-valuedness result of Donchev-Hager [14] can be deduced from Corollary 6.2.

Corollary 6.3 *Let $T : \mathbb{R}^n \rightrightarrows \mathbb{R}^n$ be a monotone set-valued mapping and $x_0 \in \text{int}$ (Dom T). Assume that T has the Aubin property around $(x_0, x_0^*) \in \text{Gr } T$ (with U and V), then $T_{U \times V}$ is locally single-valued at x_0.*

Proof According to Lemma 6.3, the localization $T_{U \times V}$ is lower semi-continuous on U. On the other hand, T being a monotone map, $T_{U \times V}$ is also monotone. Therefore, by Corollary 6.2, the map $T_{U \times V}$ is single-valued at u, for any $u \in U$.

6.4.2 Application to Normal Operator

In this subsection, we focus our attention on the normal operator of a quasiconvex function and we give sufficient conditions for this operator to be locally single-directional.

Let us fix some notations. If $g : \mathbb{R}^n \to \mathbb{R} \cup \{+\infty\}$ is a lower semi-continuous quasiconvex function and y is an element of $\text{dom } g \setminus \arg\min g$, we will denote by $\pi(y)$ the unique projection of y on the nonempty closed convex subset $\text{cl}(S^<_{g(y)})$. In what follows we will assume that the sublevel set $S_{g(y)}$ (and/or the closure of the strict sublevel set $\text{cl}(S^<_{g(y)})$) is a polyhedron that is a finite intersection of distinct half-spaces

$$H^-(a_i, b_i) = \{y \in \mathbb{R}^n : \langle a_i, y \rangle \leq b_i\}$$

for i in a finite family $I(S_{g(y)})$. The associated hyperplanes will be denoted by $H(a_i, b_i)$.

For any $y \in \mathrm{dom}\, g$, the—possibly empty—subset $I(y)$ of $I(S_{g(y)})$ stands, roughly speaking, for the set of indices of hyperplanes touching y, that is

$$I(y) = \big\{ i \in I(S_{g(y)}) \,:\, y \in H(a_i, b_i) \big\}. \tag{6.7}$$

Similarly, if $\mathrm{cl}(S_{g(y)}^<)$ is a polyhedron (say, $\mathrm{cl}(S_{g(y)}^<) = \bigcap_{i \in I(\mathrm{cl}(S_{g(y)}^<))} H^-(a_i', b_i')$), then

$$I^<(y) = \big\{ i \in I(\mathrm{cl}(S_{g(y)}^<)) \,:\, \pi(y) \in H(a_i', b_i') \big\}.$$

Note that, for any $y \in \mathrm{dom}\, g \setminus \arg\min g$, the index set $I^<(y)$ is nonempty.

Let us observe that a very simple example for which the sublevel sets are polyhedra is the case of polyhedral quasiconvex functions.

Using the above notations, at any point $y \in \mathrm{dom}\, g \setminus \arg\min g$, $N_g^a(y)$ is the normal cone at y of the adjusted sublevel set $S_g^a(y) = S_{g(y)} \cap \overline{B}(S_{g(y)}^<, \rho_y)$ with $\rho_y = \|y - \pi(y)\|$.

Proposition 6.5 *Let $g : \mathbb{R}^n \to \mathbb{R} \cup \{+\infty\}$ be a lower semi-continuous quasiconvex function and y be an element of $\mathrm{dom}\, g$ such that $y \notin \mathrm{cl}(S_{g(y)}^<)$ and $S_{g(y)}$ is a polyhedron with $I(y)$ being a singleton (say, $I(y) = \{i_0\}$). Then, the following assertions are equivalent:*

(i) N_g^a is single-directional at y;
(ii) $N_g^a(y) = \mathbb{R}^+\{y - \pi(y)\}$;
(iii) $(y - \pi(y)) \perp H(a_{i_0}, b_{i_0})$;
(iv) There exists a neighbourhood V of y such that

$$\Big[\overline{B}(S_{g(y)}^<, \rho_y) \cap V \Big] \subset \big[S_{g(y)} \cap V \big].$$

The following elementary lemma will be needed to prove Proposition 6.5.

Lemma 6.4 *Let $C \subset \mathbb{R}^n$ be a nonempty convex set, and let $y \notin \mathrm{cl}(C)$. If $\pi(y)$ is the projection of y on $\mathrm{cl}(C)$, $\rho_y = \|y - \pi(y)\|$, and*

$$H_y^- = \big\{ u \in \mathbb{R}^n : \langle y - \pi(y), u - y \rangle \le 0 \big\},$$

then

(a) $N\big(\overline{B}(C, \rho_y), y\big) = \mathbb{R}^+\{y - \pi(y)\}$;
(b) $\overline{B}(C, \rho_y) \subset H_y^-$.
(c) for all $x \in C$, $\langle y - \pi(y), x - \pi(y) \rangle \le 0$.

Proof Let us check first that for each $y, a \in \mathbb{R}^n$, one has

$$N\big(\overline{B}(a, \|y - a\|), y\big) = \mathbb{R}_+\{y - a\}. \tag{6.8}$$

Indeed,

$$y^* \in N\left(\bar{B}(a, \|y - a\|), y\right) \Leftrightarrow \forall u \in \bar{B}(0, 1), \quad \langle y^*, a + \|y - a\| u - y \rangle \leq 0$$
$$\Leftrightarrow \langle y^*, y - a \rangle \geq \sup_{u \in \bar{B}(0,1)} \langle y^*, u \rangle \|y - a\| = \|y^*\| \|y - a\|$$
$$\Leftrightarrow y^* \in \mathbb{R}^+ \{y - a\}.$$

Using relation (6.8) and the inclusion $\bar{B}\left(\pi(y), \rho_y\right) \subset \bar{B}\left(C, \rho_y\right)$, we deduce

$$N\left(\bar{B}\left(C, \rho_y\right), y\right) \subset N\left(\bar{B}\left(\pi(y), \rho_y\right), y\right) = \mathbb{R}^+ \{y - \pi(y)\}.$$

Thus, (a) follows and assertion (b) is an immediate consequence of (a). Finally, let us note that for all $x \in C$, $x + y - \pi(y) \in \bar{B}(C, \rho_y)$. Thus, (b) implies

$$\langle y - \pi(y), (x + y - \pi(y)) - y \rangle \leq 0,$$

i.e. assertion, (c).

Proof of Proposition 6.5. (i) $\Rightarrow$ (ii). Since $S_g^a(y) \subset \bar{B}(S_{g(y)}^<, \rho_y)$, we have $N\left(\bar{B}(S_{g(y)}^<, \rho_y), y\right) \subset N_g^a(y)$, and therefore, according to Lemma 6.4 (a),

$$0 \neq (y - \pi(y)) \in N_g^a(y).$$

Assertion (ii) follows from the fact that N_g^a is single-directional at y.

(ii) $\Rightarrow$ (iii). Since $S_{g(y)}^a \subset S_{g(y)} \subset H^-(a_{i_0}, b_{i_0})$, it follows that

$$a_{i_0} \in N(H^-(a_{i_0}, b_{i_0}), y) \subset N(S_{g(y)}, y) \subset N_g^a(y).$$

Thus, from ii) there exists $\lambda > 0$ such that $y - \pi(y) = \lambda a_{i_0}$, and hence, $(y - \pi(y)) \perp H(a_{i_0}, b_{i_0})$.

(iii) $\Rightarrow$ (iv). Since $I(y) = \{i_0\}$, there exists $\varepsilon > 0$ such that $B(y, \varepsilon) \cap H(a_i, b_i) = \emptyset$ for all $i \neq i_0$. From $y \in H^-(a_i, b_i)$, it follows that $B(y, \varepsilon) \subseteq H^-(a_i, b_i)$ for $i \neq i_0$. Hence,

$$S_{g(y)} \bigcap B(y, \varepsilon) = \bigcap_i \left[H^-(a_i, b_i) \cap B(y, \varepsilon)\right]$$
$$= H^-(a_{i_0}, b_{i_0}) \bigcap \left[\cap_{i \neq i_0} H^-(a_i, b_i) \cap B(y, \varepsilon)\right]$$
$$= H^-(a_{i_0}, b_{i_0}) \bigcap B(y, \varepsilon). \tag{6.9}$$

On the other hand, by Lemma 6.4 (b),

$$\bar{B}(S_{g(y)}^<, \rho_y) \subset H_y^- = H^-\left(y - \pi(y), \langle y - \pi(y), y \rangle\right).$$

But from (iii) it is clear that $H_y^- = H^-(a_{i_0}, b_{i_0})$, and thus, combining with (6.9), we obtain for $V = B(y, \varepsilon)$,

$$\left[\overline{B}(S_{g(y)}^<, \rho_y) \cap V\right] \subset \left[H^-(a_{i_0}, b_{i_0}) \cap V\right] = S_{g(y)} \cap V.$$

(iv) $\Rightarrow$ (i). From the definition of $S_g^a(y)$ and assumption (iv), we infer

$$S_g^a(y) \cap V = \overline{B}(S_{g(y)}^<, \rho_y) \cap S_{g(y)} \cap V = \overline{B}(S_{g(y)}^<, \rho_y) \cap V. \qquad (6.10)$$

Whenever $C \subseteq \mathbb{R}^n$ is convex, $x \in C$, and V is a neighbourhood of x, then $N(C \cap V, x) = N(C, x)$. We calculate the normal cone $N_g^a(y)$ by using, successively, this remark (6.10), and Lemma 6.4 (a):

$$\begin{aligned}
N_g^a(y) &= N\big(S_g^a(y), y\big) = N(S_g^a(y) \cap V, y) \\
&= N\big(\overline{B}(S_{g(y)}^<, \rho_y) \cap V, y\big) \\
&= N\big(\overline{B}(S_{g(y)}^<, \rho_y), y\big) = \mathbb{R}^+\{y - \pi(y)\}.
\end{aligned}$$

Thus, i) follows.

Let us observe that

$$I(y) = \emptyset \iff y \in \text{int}(S_{g(y)}).$$

In this case, if $y \in \text{cl}(S_{g(y)}^<)$, then $N_g^a(y) = N(S_{g(y)}^<, y)$ and N_g^a is single-directional at y if and only if $\text{card}(I^<(y)) = 1$. Finally, if $I(y) = \emptyset$ and $y \in \text{int}(S_{g(y)}) \setminus \text{cl}(S_{g(y)}^<)$, then we obtain $N_g^a(y) = N(\overline{B}(S_{g(y)}^<, \rho_y), y)$ and N_g^a is single-directional at y.

In the following proposition, we provide sufficient conditions for the nonlocal single-directionality of the normal operator.

Proposition 6.6 *Let $g : \mathbb{R}^n \to \mathbb{R} \cup \{+\infty\}$ be a lower semi-continuous quasiconvex function, and let $y \in \text{dom } g$ be such that $y \notin \text{cl}(S_{g(y)}^<)$. If $S_{g(y)}$ and $\text{cl}(S_{g(y)}^<)$ are polyhedra and $\text{card}(I(y)) \neq \text{card}(I^<(y))$, then N_g^a is not locally single-directional at y; that is for any neighbourhood V of y,*

$$\exists z \in V \text{ such that } N_g^a(z) \text{ is not single-directional.}$$

Proof Without loss of generality, we can assume that $\text{card}(I(y)) = 1$. Indeed, if $\text{card}(I(y)) > 1$, then the normal cone $N(S_{g(y)}, y)$ is not single-directional, and therefore, thanks to (6.4), this is also true for $N_g^a(y)$.

So let us now assume that $I(y) = \{i_0\}$. If $(y - \pi(y)) \not\perp H(a_{i_0}, b_{i_0})$, then the conclusion follows directly from Proposition 6.5.

So suppose now that $(y - \pi(y)) \perp H(a_{i_0}, b_{i_0})$. Since $I(y) = \{i_0\}$, there exists $\varepsilon > 0$ such that

$$B(y, \varepsilon) \cap \text{bd}(S_{g(y)}) = B(y, \varepsilon) \cap H(a_{i_0}, b_{i_0}), \qquad (6.11)$$

and thus, for any $z \in B(y, \varepsilon) \cap \mathrm{bd}(S_{g(y)})$, $I(y) = I(z)$. If ε is small enough, then we know that $z \notin \mathrm{cl}(S_{g(y)}^<)$, and hence, $g(z) = g(y)$ and $\mathrm{cl}(S_{g(y)}^<) = \mathrm{cl}(S_{g(z)}^<)$.

Let $0 < \varepsilon' \leq \varepsilon$. Suppose that for all $z \in B(y, \varepsilon') \cap \mathrm{bd}(S_{g(y)})$, $N_g^a(z)$ is single-directional. According to Proposition 6.5 and since $I(y) = I(z) = \{i_0\}$, we have

$$(z - \pi(z)) \perp H(a_{i_0}, b_{i_0}), \quad \forall z \in B(y, \varepsilon') \cap \mathrm{bd}(S_{g(y)}).$$

In particular, for some $\lambda > 0$,

$$z - \pi(z) = \lambda(y - \pi(y)). \tag{6.12}$$

Since y and z are elements of $H(a_{i_0}, b_{i_0})$, it follows that

$$\langle y - \pi(y), y - z \rangle = 0. \tag{6.13}$$

Since $\pi(z) \in \mathrm{cl}(S_{g(y)}^<)$, Lemma 6.4 (c) implies

$$\langle y - \pi(y), \pi(z) - \pi(y) \rangle \leq 0. \tag{6.14}$$

Likewise,

$$\langle z - \pi(z), \pi(y) - \pi(z) \rangle \leq 0. \tag{6.15}$$

Combining (6.12) (6.14), and (6.15), we deduce

$$\langle y - \pi(y), \pi(y) - \pi(z) \rangle = 0$$

which, together with (6.13), implies

$$\langle y - \pi(y), y - \pi(y) \rangle = \langle y - \pi(y), z - \pi(z) \rangle.$$

Therefore, in (6.12) we obtain $\lambda = 1$, so

$$y - \pi(y) = z - \pi(z) \quad \forall z \in B(y, \varepsilon') \cap H(a_{i_0}, b_{i_0}). \tag{6.16}$$

Thus, for each $z \in B(y, \varepsilon') \cap H(a_{i_0}, b_{i_0})$, $z + (\pi(y) - y) = \pi(z) \in \mathrm{bd}(\mathrm{cl}(S_{g(y)}^<))$. It follows that the translation $\pi(y) - y + B(y, \varepsilon') \cap H(a_{i_0}, b_{i_0})$ is included in $\mathrm{bd}(\mathrm{cl}(S_{g(y)}^<))$. Obviously, $\pi(y) - y + B(y, \varepsilon') = B(\pi(y), \varepsilon')$ and $\pi(y) - y + H(a_{i_0}, b_{i_0}) = H(a_{i_0}, c_{i_0})$, where $c_{i_0} = b_{i_0} + \langle a_{i_0}, \pi(y) - y \rangle$. Hence,

$$\left[B(\pi(y), \varepsilon') \cap H(a_{i_0}, c_{i_0}) \right] \subset \mathrm{bd}(\mathrm{cl}(S_{g(y)}^<))$$

and thus, $I^<(i_0) = \{i_0\}$ which contradicts $\mathrm{card}(I^<(y)) \neq \mathrm{card}(I(y)) = 1$.

In order to illustrate the above Proposition 6.5, let us consider the function g defined from $\mathbb{R}^2$ to $\mathbb{R}$ by

$$g(y_1, y_2) = \begin{cases} \max\{y_1, y_2\} & \text{if } y_1 \leq 0 \text{ and } y_2 \leq 0 \\ y_2 & \text{if } y_2 > 0 \\ 0 & \text{otherwise} \end{cases} \tag{6.17}$$

The graph and sublevel sets of the function g are represented below.

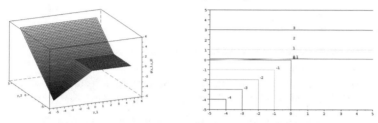

Now let us consider $y = (1, 0)$. Then, one clearly has that $y \notin \mathrm{cl}(S_{g(y)}^<) =]-\infty, 0]^2$. Moreover, $S_{g(y)} = \mathbb{R} \times]-\infty, 0]$ and $\mathrm{cl}(S_{g(y)}^<)$ are polyhedra and $1 = \mathrm{card}(I(y)) \neq \mathrm{card}(I^<(y)) = 0$. Thus according to Proposition 6.5, the adjusted normal operator N_g^a is not locally single-directional at y. Indeed, for any $\varepsilon > 0$, $N_g^a(1 + \varepsilon/2, 0) = (\mathbb{R}^+)^2$.

6.5   Dense Single-Directionality

Even if the single-directionality and single-valuedness results of Sects. 6.3 and 6.4 also hold true, respectively, in Banach spaces and Hilbert spaces, both of these sections have been restricted to $\mathbb{R}^n$, first for the sake of simplicity and second because the statements do not lose power when they are established in $\mathbb{R}^n$ instead of infinite dimensional setting. Nevertheless restricting ourself to $\mathbb{R}^n$ in this section dedicated to dense single-directionality would limit the power of the forthcoming results.

Thus in this section, $(X, \|\cdot\|)$ and X^* stand, respectively, for a Banach space and its topological dual, the latter being equipped with the weak* topology. The definitions of quasimonotonicity, single-directionality and semi-continuity given in Sect. 6.2 can be easily adapted to this infinite dimensional setting.

6.5.1   General Results

A topological space S is called *fragmentable* if there exists a metric ρ on it (equivalently, a function $\rho : S \times S \to [0, +\infty)$ with $\rho(x, y) \neq 0$, if and only if $x, y \in S$ are distinct) such that, for every $\varepsilon > 0$ and every $\emptyset \neq M \subset S$, there exists an open set $\Omega \subset S$ such that the intersection $M \cap \Omega$ is nonempty and has ρ-diameter less than

ε. An important particular example of this concept comes from Asplund spaces: if X is such, then the dual X^* provided with the weak* topology is fragmentable. Indeed if X is an Asplund space, then bounded subsets of X^* have weak* slices of diameter less than any epsilon (with respect to the metric generated with the dual norm) where for Ω one can take a weak* open halfspace, see [22, Lemma 2.18]. Then, trivially, the closed balls $\bar{B}^*(0, n)$, $n \in \mathbb{N}^*$, in dual X^*, provided with weak* topology are fragmentable (by metric coming from the norm). Then, since $X^* = \bigcup_{n\in\mathbb{N}} \bar{B}^*(0, n)$, the whole $(X^*, w*)$ is fragmentable, as a union of countably many closed fragmentable sets, see Fabian [15].

A set-valued map T is called to be

– *maximal quasimonotone* if it is quasimonotone and $T' = T$ whenever $T' : U \rightrightarrows X^*$ is quasimonotone and $T'(\cdot) \supset T(\cdot)$;
– *w^*-usco* if T is upper semi-continuous with nonempty w^*-compact values;
– *w^*-cusco* if T is upper semi-continuous with nonempty convex w^*-compact values.

Finally, let us recall that a subset V of a set U is said to be a *dense G_δ subset* of U if V is dense in U and is a countable intersection of open subsets of U.

Theorem 6.3 *Let $(X, \| \cdot \|)$ be a Banach space such that X^*, in the weak* topology, is fragmentable. Let $\emptyset \neq U \subset X$ be an open set, let $0 \notin C \subset X^*$ be a weak* compact (not necessarily convex) set, and let $Q : U \rightrightarrows X^*$ be a quasimonotone set-valued map such that $Q(x) \cap C \neq \emptyset$ for every $x \in U$. Then, there exists a dense G_δ subset $D \subset U$ such that $Q(x)$ is single-directional and lies in $[0, +\infty)C$ for every $x \in D$.*

The proof of this theorem, which is not simple, is included here for the sake of completeness. It is extracted from [8].

Proof A simple argument profiting from Zorn's lemma reveals that there exists a maximal quasimonotone mapping $Q' : U \rightrightarrows X^*$ such that $Q'(x) \supset Q(x)$ for every $x \in U$. We define $\Gamma : U \rightrightarrows C$ by $\Gamma(x) = Q'(x) \cap C$, $x \in U$. Clearly, $\Gamma(x) \neq \emptyset$ for every $x \in U$. We claim that the graph of Γ is norm$\times$weak*-closed in $U \times X^*$. Indeed, consider a net $\Gamma \ni (x_\alpha, \xi_\alpha) \longrightarrow (x, \xi) \in U \times X^*$. We shall show that (x, ξ) is "quasimonotonically related" to Q'; see [4]. So, consider any $y \in U$ such that $\langle \xi, y - x \rangle > 0$. As $\|x_\alpha - x\| \to 0$, $\xi_\alpha \to \xi$, and (ξ_α) is bounded, for all α's big enough we have that $\langle \xi_\alpha, y - x_\alpha \rangle > 0$, and so, $\inf \langle Q'(y), y - x_\alpha \rangle \geq 0$ by the quasimonotonicity of Q'. Thus, $\inf \langle Q'(y), y - x \rangle \geq 0$, and hence, $\xi \in Q'(x)$ by the maximal quasimonotonicity of Q'. Also, $\xi \in C$ as C is weak* closed. Therefore, $\xi \in \Gamma(x)$. We proved the norm$\times$weak*-closeness of Γ. The map Γ is thus also norm-to-weak* upper semi-continuous because $\Gamma(U)$ is included in the weak* compact set C. Let $\Gamma_0 : U \rightrightarrows C$ be a minimal norm-to-weak* usco mapping such that $\Gamma_0(\cdot) \subset \Gamma(\cdot)$; its existence is again guaranteed by Zorn's lemma.

Now, we are ready to apply [15, Proposition 5.1.11]. It yields a dense G_δ subset $D \subset U$ such that $\Gamma_0(x)$ is a singleton for every $x \in D$.

Fix any such $x \in D$. Then, $\Gamma_0(x) \neq \{0\}$ since $0 \notin C$. It remains to check the single-directionality of Q at x, that is that $Q(x) \subset [0, +\infty)\Gamma_0(x)$. Assume that there exists a $\xi \in Q(x) \setminus [0, +\infty)\Gamma_0(x)$. A simple indirect argument yields a $0 < \delta < \|\Gamma_0(x)\|$ so small that $\xi \notin [0, +\infty)(\Gamma_0(x) + \delta B_{X^*})$, where $B_{X^*} := \{x^* \in X^* : \|x^*\| \leq 1\}$. Find $v \in X$ such that $\|v\| = 1$ and $\langle \Gamma_0(x), v \rangle > \delta$. We shall check that the cone $[0, +\infty)(\Gamma_0(x) + \delta B_{X^*})$ is weak* closed. So, consider a net $(x_\alpha^*)_{\alpha \in A}$ in it that weak* converges to some $x^* \in X^*$. For every $\alpha \in A$ find $\lambda_\alpha \geq 0$ and $z_\alpha^* \in B_{X^*}$ such that $x_\alpha^* = \lambda_\alpha(\Gamma_0(x) + \delta z_\alpha^*)$. Then,

$$1 + \langle x^*, v \rangle > \langle x_\alpha^*, v \rangle = \lambda_\alpha \langle \Gamma_0(x), v \rangle + \lambda_\alpha \delta \langle z_\alpha^*, v \rangle \geq \lambda_\alpha \big(\langle \Gamma_0(x), v \rangle - \delta \big)$$

for all $\alpha \in A$ big enough. Hence, all cluster points of the net $(\lambda_\alpha)_{\alpha \in A}$ are bounded by $\big(1 + \langle x^*, v \rangle\big)/(\langle \Gamma_0(x), v \rangle - \delta)$ $(< +\infty)$. Finally, let $(\lambda, z^*) \in [0, +\infty) \times B_{X^*}$ be a weak* cluster point of the net $(\lambda_\alpha, z_\alpha^*)_{\alpha \in A}$. Then, $x^* = \lambda(\Gamma_0(x) + \delta z^*) \in [0, +\infty)(\Gamma_0(x) + \delta B_{X^*})$.

Now, we are ready to use the Hahn-Banach separation theorem for the point ξ and the weak* closed cone $[0, +\infty)(\Gamma_0(x) + \delta B_{X^*})$. It yields $0 \neq u \in X$ so that $\langle \xi, u \rangle > \sup \langle [0, +\infty)(\Gamma_0(x) + \delta B_{X^*}), u \rangle$. Clearly, the right-hand side here is 0. Thus,

$$\langle \xi, u \rangle > 0 \geq \sup \langle \Gamma_0(x) + \delta B_{X^*}, u \rangle = \langle \Gamma_0(x), u \rangle + \delta\|u\| > \langle \Gamma_0(x), u \rangle.$$

Hence, $\Gamma_0(x)$ lies in the (weak* open) set $\{\langle \cdot, u \rangle < 0\}$. Then, the norm-to-weak*-upper semi-continuity of Γ_0 yields a $t > 0$ such that $x + tu \in U$ and $\Gamma_0(x + tu) \subset \{\langle \cdot, u \rangle < 0\}$. Pick some $\eta \in \Gamma_0(x + tu)$; then $\langle \eta, u \rangle < 0$. On the other hand, as $\sup \langle Q'(x), x + tu - x \rangle \geq t\langle \xi, u \rangle > 0$, the quasimonotonicity of Q' yields that

$$t\langle \eta, u \rangle = \langle \eta, x + tu - x \rangle \geq \inf \langle Q'(x + tu), x + tu - x \rangle \geq 0.$$

But $\langle \eta, u \rangle < 0$, a contradiction. We proved that $Q(x) \subset [0, +\infty)\Gamma_0(x)$.

Note that in [4], the authors defined a different concept of maximality for quasimonotone mappings. Indeed, a multivalued mapping $T : X \rightrightarrows X^*$ is said to be maximal quasimonotone (in the sense of [4]) if T is quasimonotone on its domain and for any quasimonotone mapping $\widehat{T} : X \rightrightarrows X^*$ such that $T(x) \subseteq \text{cone } \widehat{T}(x)$ for all $x \in \text{Edom } T$ one has cone $T(x) = \text{cone } \widehat{T}(x)$ for all $x \in \text{Edom } T$ and $\text{Edom } T = \text{Edom } \widehat{T}$ where the effective domain, Edom $\widehat{T}$, is defined as

$$\text{Edom } T := \big\{x \in \text{Dom } T : T(x) \neq \{0\}\big\}.$$

This latter definition of maximality could appear to be rather complicated when comparing with the classical "inclusion-type" one used in the proof of Theorem 6.3. Actually, it is important to note that, in the context of Theorem 6.3, both coincide. Indeed, in Theorem 6.3, it is assumed that $0 \notin C$ and for any $x \in U$, $Q(x) \cap C$ is nonempty, which immediately implies that U is included in the effective domain Edom Q. More precisely, the relationship on Edom Q between both concepts of maximality is

$$\left.\begin{array}{l} Q \text{ is maximal quasimonotone} \\ \text{in the sense of [8]} \end{array}\right\} \Leftrightarrow \left\{\begin{array}{l} cone(Q) \text{ is maximal quasimonotone} \\ \text{in the inclusion sense.} \end{array}\right.$$

Let us now consider the case cone-valued maps.

Corollary 6.4 *Let X be a Banach space such that X^*, in the weak* topology, is fragmentable. Let $\emptyset \neq U \subset X$ be an open set. Let $Q : U \rightrightarrows X^*$ be a quasimonotone and cone-valued map. Assume that there exists $s \in X$ such that*

$$Q(x) \subset \{x^* \in X^* : \|x^*\| \leq \langle x^*, s \rangle\}, \quad \text{for every } x \in U.$$

Then, there exists a dense G_δ subset $D \subset U$ such that $Q(x)$ is single-directional for every $x \in D$.

Proof Apply Theorem 6.3 for the set $C := \{\|\cdot\| \leq \langle \cdot, s \rangle = 1\}$.

6.5.2 Dense Single-Directional Property: The Case of the Normal Operator

Our aim in this section is to consider the case of quasiconvex functions and to provide some sufficient conditions for the associated normal operator to be single-directional on a dense subset of its domain. Based on Theorem 6.3, we provide below the generic single-directionality of the adjusted normal operator.

Theorem 6.4 *Let $(X, \|\cdot\|)$ be a Banach space, let $\emptyset \neq V \subset X$ be an open set, let $g : V \to \mathbb{R}$ be a quasiconvex lower semi-continuous and solid function, and let $x \in V$ be such that $\operatorname{argmin}_V g$ is a subset of $S_{g(x)}^<$ with $\operatorname{argmin}_V g \neq S_{g(x)}^<$. If X^*, with the weak* topology, is fragmentable, then there exists an open set $x \in U \subset V$ such that the adjusted normal operator $N_g^a : U \rightrightarrows X^*$ is single-directional at any point of a dense G_δ subset of U.*

Proof Find $x_1 \in V$ such that $\inf_V g < g(x_1) < g(x)$. Since g is solid, there are $y \in V$ and $\delta > 0$ such that $y + 2\delta B_X \subset S_g^a(x_1)$. From the lower semi-continuity of g, find $\gamma \in (0, \delta)$ so small that $g(x') \geq g(x_1)$ whenever $x' \in \{x' \in V : \|x' - x\| < \gamma\} =: U$. Then for every $x' \in U$, we have that $y + 2\delta B_X \subset S_g^a(x')$, and so, considering any $0 \neq x^* \in N_g^a(x')$, we can estimate

$$\langle x^*, y \rangle + 2\delta \|x^*\| = \sup \langle x^*, y + 2\delta B_X \rangle \leq \sup \langle x^*, S_g^a(x') \rangle \leq \langle x^*, x' \rangle < \langle x^*, x \rangle + \delta \|x^*\|,$$

and hence, $\delta \|x^*\| < \langle x^*, x - y \rangle$. We proved that

$$N_g^a(x') \setminus \{0\} \subset \left\{ x^* \in X^* : \delta \|x^*\| < \langle x^*, x - y \rangle \right\}, \quad \text{for every } x' \in U.$$

Let C be the weak* closure of the set $\{x^* \in X^* : \|x^*\| = 1 \text{ and } \langle x^*, x - y \rangle > \delta\}$; then $0 \notin C$. Now the conclusion follows by applying Theorem 6.3.

To illustrate the situation described in Theorem 6.4, let us consider the continuous quasiconvex function defined in (6.17). As it can be easily verified, the above result can be applied to any $y \in \mathbb{R}^2$ and the set of nonsingle-directionality of N_g^a is

$$\mathcal{D} = \{(y_1, y_1) \, : \, y_1 \leq 0\} \cup \{(y_1, 0) \, : \, y_1 > 0\}.$$

References

1. Aussel, D., Corvellec, J.-N., Lassonde, M.: Subdifferential characterization of quasiconvexity and convexity. J. Convex Anal. **1**, 195–201 (1994)
2. Aussel, D., Cotrina, J.: Stability of quasimonotone variational inequality under sign-continuity. J. Optim. Theory Appl. **158**, 653–667 (2013)
3. Aussel, D., Dutta, J.: Generalized Nash equilibrium problem, variational inequality and quasiconvexity. Oper. Res. Lett. **36**(4), 461–464 (2008). [Addendum in Oper. Res. Lett. **42**, 398 (2014)]
4. Aussel, D., Eberhard, A.: Maximal quasimonotonicity and dense single-directional properties of quasimonotone operators. Math. Program. **139**, 27–54 (2013)
5. Aussel, D., García, Y.: On Extensions of Kenderov's Single-Valuedness Result for Monotone Maps and Quasimonotone Maps. SIAM J. Optim. **24–2**, 702–713 (2014)
6. Aussel, D., García, Y., Hadjisavvas, N.: Single-directional property of multivalued maps and variational systems. SIAM J. Optim. **20**, 1274–1285 (2009)
7. Aussel, D., Hadjisavvas, N.: Adjusted sublevel sets, normal operator, and quasi-convex programming. SIAM J. Optim. **16**, 358–367 (2005)
8. Aussel, D., Fabian, M.: Single-directional properties of quasi-monotone operators. Set-valued Var. Anal. **21**, 617–626 (2013)
9. Aussel, D., Ye, J.: Quasi-convex programming with starshaped constraint region and application to quasi-convex MPEC. Optimization **55**, 433–457 (2006)
10. Borwein, J., Lewis, A.: Convex Analysis and Nonlinear Optimization. Theory and examples, Second edition, CMS Books in Mathematics/Ouvrages de Mathématiques de la SMC, 3, Springer, New York (2006)
11. Borde, J., Crouzeix, J.-P.: Continuity properties of the normal cone to the level sets of a quasiconvex function. J. Optim. Theory Appl. **66**, 415–429 (1990)
12. Christensen, J.P.R.: Theorems of Namioka and R.E. Johnson type for upper-semicontinuous and compact-valued set-valued mappings. Proc. Amer. Math. Soc. **86**, 649–655 (1982)
13. Christensen, J.P.R., Kenderov, P.S.: Dense strong continuity of mappings and the Radon-Nikodým property. Math. Scand. **54**, 70–78 (1984)
14. Dontchev, A.L., Hager, W.: Implicit functions, Lipschitz maps, and stability in optimization. Math. Oper. Res. **19**, 753–768 (1994)
15. Fabian, M.J.: Gâteaux Differentiability of Convex Functions and Topology—Weak Asplund Spaces. Wiley, New York (1997)
16. Daniilidis, A., Hadjisavvas, N., Martinez-Legaz, J.E.: An appropriate subdifferential for quasiconvex functions. SIAM J. Optim **12**, 407–420 (2001)

17. Hadjisavvas, N.: Continuity and maximality properties of pseudomonotone operators. J. Convex Anal. **10**, 459–469 (2003)
18. Kenderov, P.: Semi-continuity of set-valued monotone mappings. Fund. Math. **LXXXVIII**, 61–69 (1975)
19. Kenderov, P.: Dense strong continuity of pointwise continuous mappings. Pacific J. Math. **89**, 111–130 (1980)
20. Kenderov, P., Moors, W., Revalski, J.: Dense continuity and selections of set-valued mappings. Serdica Math. J. **24**, 49–72 (1998)
21. Levy, A., Poliquin, R.: Characterizing the single-valuedness of multifunctions. Set-Valued Anal. **5**, 351–364 (1997)
22. Phelps, R.R.: Convex Functions, Monotone Operators, and Differentiability, Lecture Notes in Math, vol. 1364, 2nd edn. Springer Verlag, Berlin (1993)

Printed in the United States
By Bookmasters